Interfacing Biological Equipment with Microcomputers:
using the BBC model B microcomputer

John F. Brown, B.Sc., Ph.D.
Head of Combined Studies,
Lancashire Polytechnic (Preston)

Robert J. Strettle, B.Sc., Ph.D.
Senior Lecturer in Physiology and Pharmacology,
Lancashire Polytechnic (Preston)

Colin J. Sutherland, B.Sc., Ph.D.
Senior Lecturer in Physiology,
Lancashire Polytechnic (Preston)

Edward Arnold

First published in Great Britain 1986 by
Edward Arnold (Publishers) Ltd, 41 Bedford Square, London WC1B 3DQ

Edward Arnold (Australia) Pty Ltd, 80 Waverley Road, Caulfield East,
Victoria 3145, Australia

Edward Arnold, 3 East Read Street, Baltimore, Maryland 21202, U.S.A.

British Library Cataloguing in Publication Data

Brown, John F.
  Interfacing biological equipment with
  microcomputers: using the BBC model B
  microcomputer.
  1. Computer interfaces   2. BBC Microcomputer
  I. Title   II. Strettle, Robert J.
  III. Sutherland, Colin J.
  004.6'16   TK7887.5
  ISBN 0-7131-2937-9

While every precaution has been taken in the preparation of this book,
the authors and publishers assume no liability for any damages resulting
from the use of the information contained in this book.

Text set in 9/11 pt Univers
by Colset Pte Ltd, Singapore
Printed and bound in Great Britain by
Billing and Sons Ltd, Worcester.

Preface

This book is intended for students, lecturers and researchers in the biological and medical sciences. It is designed for both experienced programmers who wish to develop into interfacing and for the complete beginner who wishes to learn useful programming.

In our opinion microcomputers can play a number of important roles in the biological and medical sciences, but we believe that their most outstanding contribution lies in equipment control and the on line sampling, storage and analysis of data. This requires communication to occur between the experimental equipment and the computer, a technique which is known as interfacing. This book attempts to remove the mystique of interfacing and show that the simplest approach is generally the most appropriate for the life sciences.

Our approach has been developed from running a number of short courses at Lancashire Polytechnic and elsewhere on interfacing biological equipment to computers. From this experience of teaching interfacing we have found that even complete beginners to computing can be writing quite sophisticated programs within a few days. We have attempted to present the information in what we consider to be a logical sequence, however it should be possible to read any chapter or chapters in isolation as, wherever appropriate, cross-references to related information have been provided. We would recommend that you should read the overview presented in Chapter 1 before you attempt any other chapters if the concept of practical interfacing is new to you.

We have found that our progress in computing developments has been greatly assisted by many factors, two of the most important ones being:

(i) ensuring that we are developing our programs/interfacing around a particular experiment;

(ii) using a signal simulator unit for testing programs as they develop. The simulator allows us to mimic the biological signal which we would expect to obtain, without requiring the whole experiment to be performed at the development stage.

With regard to the hardware requirements for interfacing (the electronic components and leads etc.) the strategy we have adopted is to suggest particular suppliers of relevant hardware whilst at the same time outlining the specific requirements for particular situations.

All the programs listed in this book are available on disc from the Business and Industrial Centre, Lancashire Polytechnic, Preston PR1 2TQ.

The authors would like to thank H.M.I. Ken Thomas for recognizing the importance of, and for his confidence in, our work. Mike Peek of the School of Electronic and Electrical Engineering of Lancashire Polytechnic was always around late in the evening and helped to explain the complexity of the 1 MHz bus. Finally we would like to thank our students who via their enthusiasm for their computing projects have taught us some new approaches and a lot about the problems of computer interfacing.

Lancashire   John Brown
1986   Rob Strettle
                                                             Colin Sutherland

Acknowledgement

We are grateful to The Cambridge Microcomputer Centre, 153–154 East Road, Cambridge, UK for permission to reproduce the information contained in Figures 2.2, 7.1 and 7.2.

Contents

1. An Overview of Interfacing Equipment with the BBC Microcomputer

One of the most important applications of microcomputers in biology is that they can be linked up to laboratory equipment. This is known as interfacing and allows analysis of experimental results and equipment control to be performed automatically.

Several discrete stages are involved in interfacing laboratory equipment with microcomputers and these are shown in Fig. 1.1.

Each of these stages will now be considered in turn.

The Equipment

Most laboratory equipment is capable of providing, or can be made to provide, an electrical output corresponding to the events being measured.

Unfortunately the maximum voltage output from laboratory equipment can be extremely variable – from as little as a few millivolts up to several volts. In order to connect equipment to the BBC microcomputer the voltage output from the equipment must fall within a well defined range. For example the analogue input port ('analogue in' port) on the rear of the

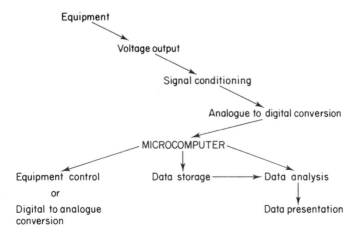

Fig. 1.1 Stages involved in interfacing.

BBC needs an input voltage in the range 0–1.8 volts. (The input requirements of the User port are dealt with in Chapter 7.)

Signal conditioning

The output voltage of laboratory equipment will probably not fall within the voltage requirements for input to, for instance, the 'analogue in' port on the BBC. Therefore one of the first things to be done is to manipulate the voltage output of the equipment so that it falls within the necessary limits. The signal may need to be amplified or attenuated, a process known as 'signal conditioning'. Signal conditioning is dealt with in Chapter 2.

Analogue to digital conversion

Microcomputers do not understand gradually changing or steady (analogue) voltages, they only understand numbers. Therefore the analogue input voltage must be converted (digitized) into a number the computer can recognize. This is the role of the analogue to digital (A–D) converter.

A–D converters vary in the number they produce for a given analogue input voltage. An 8 bit A–D converter will give a number in the range of $0-2^8$ i.e. 0–255 over the input voltage range (see Chapter 3 and Appendix 3 if you do not understand bits and bytes). A 12 bit converter will provide a number between $0-2^{12}$ (0–4095) for the same voltage range. A range of commercially available A–D converters are available which connect into the User port or the 1 MHz bus on the underside of the BBC.

Fortunately for Biologists the BBC has four A–D converters fitted into the rear of the machine and labelled 'analogue in'. These are ideal for the majority of Biological applications. Each A–D converter is 12 bit, i.e. it produces a number in the range $0-2^{12}$ (0–4095). The manufacturers of the BBC have arranged for this digital number to be multiplied by 16 so that the final digital number entering the computer will be in the range 0–65520.

Thus the 'analogue in' ports require an input voltage between 0 and 1.8 volts which will be converted into a number between 0 and 65520 (see Table 1.1).

Table 1.1

Input volts	Digital number entering computer
0	0
0.25	9100
0.5	18200

Table 1.1   *contd*

Input volts	Digital number entering computer
0.75	27300
1.0	36400
1.25	45500
1.5	54600
1.75	63700
1.8	65520

Relationship between the input voltage to the 'analogue in' port and the digital number entering the BBC microcomputer.

Entering the digital number into the computer

The method employed to enter the digital number into the computer varies according to whether the input is via the 'analogue in' port, the User port or the 1 MHz bus. Input via the 'analogue in' port will be dealt with here (see Chapter 7 for User port and 1 MHz bus).

The model B BBC microcomputer has four A–D converter input channels numbered 1, 2, 3 and 4 (though to confuse the issue the wiring diagram in the BBC User Guide labels them 0, 1, 2 and 3).

To call in a digital number from a particular channel the ADVAL instruction is used (*A* to *D* converter *VAL*ue) and the channel to be used is signified, i.e. ADVAL(1) will call in a number from channel 1.

Channels can be sampled every 10 milliseconds or more slowly if required. If more than one channel is in use it requires approximately 10 milliseconds to sample each channel in sequence.

The following simple program gives a basic input routine.

```
10 CLS
20 FOR N = 1 TO 10
30 LET B = ADVAL(1)
40 PRINT B
50 NEXT N
```

This program will take 10 samples into a variable B, each at approximately 10 millisecond intervals (the PRINT statement will take up some time also). The values of B will be printed on the screen.

For many biological purposes a sample every 10 milliseconds is a little excessive. The sampling rate may be controlled to the user's requirements by using the computer's own internal clock. This clock counts in

10 millisecond intervals and may be accessed using the TIME, INKEY or INKEY$ instructions. Therefore the sampling rate can be delayed by one second by the use of lines 50 and 60 in the following program.

```
10 CLS
20 FOR N = 1 TO 10
30 LET B = ADVAL(1)
40 PRINT B
50 TIME = 0
60 REPEAT UNTIL TIME = 100
70 NEXT N
```

Alternatively lines 50 and 60 could be replaced with either C = INKEY(100) or C$ = INKEY$(100).

Data manipulation and analysis

So far samples have been taken into the computer and entered into a variable – B. However B is updated each time a sample is taken and thus all earlier data is lost.

Data manipulation requires the use of the incoming information. For example, the incoming data may be plotted on the screen in the same way as it is normally plotted on a chart recorder. The following program would allow this to be done.

```
10 MODE 0
20 FOR N = 1 TO 1000
30 LET B = ADVAL(1)
40 DRAW N,B/64
50 NEXT N
```

This program draws across the screen using as its X co-ordinate the current value of N in the loop and as its Y co-ordinate the incoming value of B (see Chapter 5 for details of using the DRAW instruction). B had to be divided by 64 since the peak digital number from the A–D converter (65520) vastly exceeds the number of co-ordinates on the Y axis of the monitor screen (0 to 1023, see Chapter 5).

In a similar fashion the incoming data could have been used to plot a graph or histogram. Again however, although a picture appears on screen, the information has been lost. If further analysis on the data is to be performed it will need to be stored in memory or on disc. Data analysis is covered in detail in Chapter 6 and data storage is covered in Chapters 3 and 4.

Output from the computer

The BBC microcomputer has a number of output facilities. These include the User and Printer ports and the 1 MHz bus.

Two types of output control are available:
(a) on/off switching to control equipment such as stepper motors, mains switches etc.;
(b) a variable analogue output voltage which requires a separate digital to analogue (D–A) converter to be connected to the computer.

These output control facilities of the BBC are dealt with in Chapter 7.

2. Signal Input into the BBC Microcomputer

It is likely that one of the first questions to be asked is 'Will I be able to link my .... (please insert an appropriate word such as spectrophotometer, pH meter, transducer etc.) to the computer?'. In the majority of situations the answer to this question is almost certainly yes, especially if the device in question can be connected to a pen recorder. If the piece of equipment does not have an obvious electrical output do not despair, it may still be possible to link it to a computer. A few ideas on this topic are presented at the end of this chapter.

Selection of the input port

The first question which needs to be considered is 'To which input port should the equipment be connected?' To a certain extent this depends upon the nature of the electrical signal, which may be either an analogue (a continuously variable voltage) or digital (a signal which can only be on or off) signal. In the majority of cases the signal will probably be an analogue signal so this will be the main consideration here (see Chapter 7 for digital input).

There are two major mechanisms for the connection of input devices to the BBC microcomputer. These are via:

(i) The 'analogue in' port which can be found on the rear of the microcomputer;

(ii) Either the User port or the 1 MHz bus which can both be found on the underside of the microcomputer.

An example of how such connections might be made is illustrated in Fig. 2.1.

From Fig. 2.1 it is evident that to connect to the 'analogue in' port you need only a signal conditioning unit (Palmer Bioscience supply an appropriate unit), whereas to connect to either the User port or the 1 MHz bus you require two units, a signal conditioning unit and an A–D converter. There are a range of suitable interface devices available. Some of these contain both a signal conditioning unit and an A–D converter (examples being the Palmer Bioscience and Phillip Harris interface units) and some contain only A–D converters (an example being the Interbeeb Unit supplied by Griffin & George).

In the authors' experience, input via the 'analogue in' port is probably the most appropriate in the majority of cases for the following reasons.

(a)

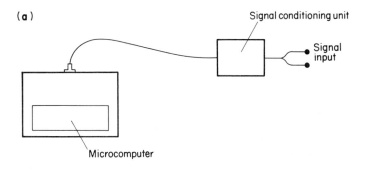

(b)

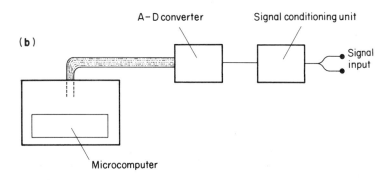

Fig. 2.1 Input connections to the BBC microcomputer via **(a)** the 'analogue in' port, and **(b)** the User port or 1 MHz bus.

(i)   The cost of external interface components is considerably less than when using the User port or 1 MHz bus, as you are using the computer's own A–D converter.

(ii) Programming is easier to understand when using the 'analogue in' port, as there is no need for an understanding of indirection operators (see Chapter 3).

(iii) The accuracy of determinations is greater when using the 'analogue in' port as the A–D converter produces numbers in the range 0–4095 (the computer then converts the numbers to cover the range 0–65520). Usually when using the User port or 1 MHz bus the analogue to digital converters only produce numbers in the range 0–255 for a similar voltage input.

Situations when input via the User port or 1 MHz bus may be more appropriate can be envisaged, a few examples being:

(i)   If more than the 4 input channels that are available on the 'analogue in' port are required;

(ii) If readings are required at a faster rate than one every 10 milliseconds,

since this is the fastest sampling rate of the 'analogue in' port;
(iii) If commercial software based upon the User port or 1 MHz bus is to be used;
(iv) If purchase of a 'complete' interface package which is suitable for use as both an input and output interface is contemplated.

Thus in the majority of examples of programming quoted in this book the input to the computer will be via the 'analogue in' port. It will therefore be necessary to connect the piece of equipment to the computer using a signal conditioning unit as shown in Fig. 2.1a.

Simulating biological signals

When developing interfacing programs it may be very inconvenient to have to connect a computer to a series of pieces of equipment, and to have to run a complete experiment to test a part or all of the program. One alternative is to use a device which can mimic the expected signal. Lancashire Polytechnic Business and Industrial Centre market such a signal simulator unit. This is a very simple device with a single variable control which allows the voltage input into the 'analogue in' port to be varied over the range of 0–1.8 volts thus enabling manual simulation of a range of biological signals.

Use of the 'analogue in' port

The programming of the 'analogue in' port has already been described in Chapter 1. In this section the characteristics of the 'analogue in' port are described.

The 'analogue in' port is a 4 channel input port which allows four independent analogue voltages, in the range 0–1.8 volts, to be fed into the computer. The arrangement of the 4 input channels is shown in Fig. 2.2.

It is possible to read a digital value (after analogue to digital conversion) from any one of the 4 channels using the ADVAL instruction followed by the channel number in brackets, i.e.

    LET B = ADVAL(1)

will read a value from channel 1 into a variable B;

    LET C = ADVAL(4)

will read a value from channel 4 into a variable C.

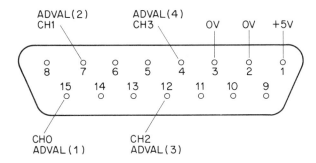

Fig. 2.2 Input connections for the 'analogue in' port viewed from the rear of the BBC microcomputer. The nomenclature is as described in the BBC User Guide and please note that the channel numbers disagree with the appropriate ADVAL instruction number.

A program containing the ADVAL instruction can sample the 'analogue in' port no faster than once every 10 milliseconds. However we have found that the input is such that the required channel may not have updated its sample after this 10 millisecond interval. Consequently the same number can be read from the 'analogue in' port on more than one visit although the input signal is changing. This problem can be overcome if all 4 channels are not in use, by using the *FX16,X instruction. This statement disables all channels with a number greater than X, i.e. *FX16,1 disables channels 2, 3 and 4 and ensures accurate recording on channel 1.

Unfortunately the 'analogue in' channels only respond to voltages in the range 0–1.8 volts and thus if a piece of equipment which has either a low voltage output (less than 0.5 volts) or a high voltage output (above 1.8 volts) is to be connected it will be necessary to either amplify or attenuate the signal respectively.

Probably the easiest way to overcome this problem is to use a suitable signal conditioning unit (such as the signal conditioning unit available from Palmer Bioscience). If you are lucky enough to be using a piece of equipment which has a voltage output in the correct range it is possible to connect the equipment directly (via a suitable lead and plug) to the 'analogue in' port. However there is a danger that the computer may be damaged by a voltage surge when switching the equipment on or upon changing sensitivity ranges. Thus for direct input to the 'analogue in' port it is necessary to guard the computer against damage and it may be convenient to use a signal conditioning unit. A simple protection circuit, as illustrated in Appendix 1, should be used if it is necessary to input a signal directly into the computer.

Adaptation of equipment for computer interfacing

Probably the majority of equipment which does not have an obvious voltage output will fall into one of two categories. The first category is equipment which is using electrical signals to cause deflection of a meter, galvanometer or similar display. In such pieces of equipment it is necessary to connect a pair of leads across the output to the meter and use these for the signal input. However, it is essential to be extremely careful as there may be danger of an electric shock if certain components are touched or a wrong connection is made, or alternatively the equipment may be damaged. Hence the maxim is – if there is any uncertainty, **don't**. Go and consult an expert.

The second category is equipment (or scientific experiments) with no evidence of a suitable electrical signal existing. In this case a transducer must be used to convert the particular event into an electrical signal.

A whole text could be devoted to the use of transducers in biology, however we can only hope to make the reader aware of some of the possibilities (Table 2.1). An example of how equipment with no electrical output can be made to give one and be interfaced is given in detail in Chapter 8.

Table 2.1

Parameter to be transduced	Transducer
Linear movement	Isotonic transducer
Rotational movement	A potentiometer (single or 10 turn)
Force/pressure	Strain gauge
Ion concentration	(a) pH or ion selective electrode
	(b) Conductivity
Flow	(a) Electromagnetic transducer
	(b) Doppler flowmeter
	(c) Electrospirometer flowhead
Light intensity	Photoelectric cell
Liquid level	(a) Conductivity
	(b) Strain gauge

A list of possible transducers of interest in biology.

The approach to take depends upon the particular situation. If it is essential to use a particular piece of equipment (for example a spirometer or plate counter or animal activity monitor) then the most appropriate way to produce an electrical output of the parameter being measured must be selected. Alternatively for some experiments it may be better to change the approach by designing the apparatus around the particular transducer; e.g. if a manometer is being used to measure pressure, replace the manometer with a pressure transducer. With a bit of luck and ingenuity it is usually possible to interface most experiments to a computer.

3. Data Storage in Memory

One great advantage that microcomputers have over other types of recording device is their ability to store data so that it may be manipulated and reused. Several methods are available for the storage of data in the computer's own memory.

Storage using variables and arrays

The simplest method and one that all programmers will be familiar with is the use of variables, i.e.

LET V = 200

tells the computer to store the number 200 in a variable called V somewhere deep in the computer's memory. Variables may be given a wide range of names but their use is limited in interfacing since they are more appropriate for storing single values rather than sets of data. They are however a useful source of temporary storage which is updated as the program proceeds (see their use in the ADVAL statement in Chapters 1 and 2).

A more appropriate data storage device is the variable array. An array is a long list of variables each capable of storing a piece of data. An array may be of any length, providing there is sufficient memory space, but the computer needs advanced warning so that it can allocate sufficient memory to your array. This is achieved with the DIM (dimension) instruction, i.e.

DIM C(500)

tells the computer to reserve memory for 500 elements of data which are to be stored in the variables C(1), C(2), C(3) to C(500). (In practice the computer would allocate 501 locations or elements since C(0) is also valid.) Using an array the variable C( ) is easily updated so that each incoming element of data is stored in a separate variable, i.e.

```
10  DIM C(1000)
20  FOR N = 1 TO 1000
30  LET C(N) = ADVAL(1)
40  NEXT N
```

would take in 1000 samples through A–D converter channel 1 and place these values in C(1) to C(1000). The loop value of N is used in line 30 to update the element in the array of C( ).

Arrays provide a simple means of storing incoming data which can easily be retrieved from memory. For example the above program could be expanded as follows:

```
50  MODE 0
60  FOR N = 1 TO 1000
70  DRAW N,C(N)/64
80  NEXT N
```

which would plot the data stored in the array across the screen.

The disadvantage of arrays is that they use a lot of memory and there are more economical ways of using the computer's memory.

Storing data directly in memory

Bits, bytes and computer memory

Because arrays are wasteful of memory a more appropriate method of data storage is to place the data directly in the computer's memory. This is a very easy procedure but to get the most out of it requires a little knowledge of the way in which the computer stores data.

Most computers (including the BBC) store numbers in binary form, i.e. a number system based solely on the digits 0 and 1. These binary digits 0 and 1 are called bits. A single bit has a limited use since it can only store the numbers 0 and 1. To overcome this problem the BBC uses a group of 8 bits known as a byte. Each bit can be either 'on' or 'off', i.e. it may have the number '1' or '0'. Each bit within a byte represents one in the series of 2 to the power of (0, 1, 2, 3, etc.). For example bit 6 represents '0' when switched off and 2^6 i.e. 64 when switched on. Table 3.1 shows a few examples of number storage within one byte.

Table 3.1

Bit	7	6	5	4	3	2	1	0
Represents	2^7	2^6	2^5	2^4	2^3	2^2	2^1	2^0
	128	64	32	16	8	4	2	1
A	0	1	0	0	0	0	0	1 = 65
B	0	0	0	0	0	0	0	0 = 0
C	1	1	1	1	1	1	1	1 = 255

Illustrating storage of numbers in a binary form in one (eight bit) byte.

Thus one byte (8 bits) can store any number between 0 (example B) and 255 (example C). Example A shows that the binary code 01000001 means (1 × 64) + (1 × 1) = 65. Appendix 3 contains programs to illustrate the conversion between 8 bit binary and decimal notation.

On the BBC each memory location is 1 byte (i.e. 8 bits) and can therefore store a number between 0 and 255. It should now be obvious that if a number greater than 255 is to be stored, more than one memory location will be needed in which to do so. For example the variables and arrays discussed above each require **five** memory locations (5 bytes) to store a single data element.

Memory map

The model B BBC microcomputer has over 32000 memory locations (bytes) of Random Access Memory (RAM). This is the part of memory that can be addressed directly and can store data. These memory locations are numbered from 0 to 32767. This sounds like an enormous amount of

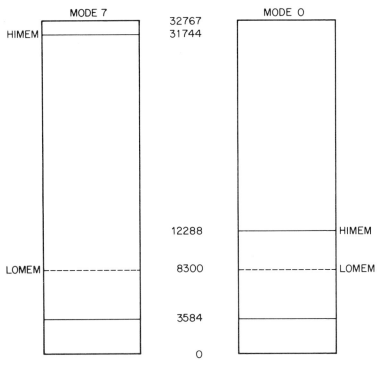

Fig. 3.1   Memory maps (RAM) showing available storage space in MODE 7 and MODE 0.

memory and few users, at the outset, would dream of using more memory than this.

Unfortunately these 32000 bytes of memory are never all available to the user since other parts of the software impinge on the memory (see Fig. 3.1).

The first 3584 memory locations are allocated to the system variables. These are general housekeeping routines linking with the computer's operating system and should not be confused with the memory locations used for storing the variables in a program.

A program has to be stored in RAM and this is loaded in from location 3584 upwards. Therefore the longer the program, the less available memory for data storage.

The computer graphics and monitor screen memory are also stored in RAM. To make things interesting this is stored from location 32767 downwards.

Screen memory in MODE 7 (Teletext display) is loaded down to location 31744. Using graphic modes or other text modes uses more memory and the higher the resolution of the graphics, the more memory required to store that screen information (Table 3.2).

Table 3.2

MODE	Graphics/Text loaded from memory location 32767 down to
0	12288
1	12288
2	12288
3	16384
4	22528
5	22528
6	24576
7	31744

Use of memory to store graphics and text in various modes.

Thus in MODE 0, with an average sized program (loaded up to memory location 8300) the 32000 locations available for data storage have been suddenly reduced to about 4000 (12288 − 8300). This may sound terribly confusing and you may well ask how you are ever going to know exactly where you can put your data. Fortunately this is easily done by using the instructions HIMEM and LOMEM. Thus the statement

    PRINT LOMEM

will indicate where the top end of your program finishes and the first free memory location starts. Note that as you continue to write your program the position of LOMEM will increase.

PRINT HIMEM

will indicate the uppermost memory location you can use.

Addressing memory locations directly

At last how much (or little) memory is available for data storage is known. The methods available for placing data into memory and withdrawing data from memory now need to be examined. This is achieved on the BBC by the use of indirection operators. The three indirection operators are shown in Table 3.3.

Table 3.3

Indirection operator	Name	Action
?	query	addresses 1 memory location (byte)
!	pling	addresses 4 memory locations (bytes)
$	dollar	addresses up to 255 memory locations (bytes)

The indirection operators.

The indirection operators ? and ! are the most appropriate for interfacing work and the discussion here will be restricted to them.

The indirection operators can be used to place data in or retrieve it from specified memory locations. The BBC User Guide suggests that memory can only be addressed using hexadecimal notation. This is not the case. If the hexadecimal notation is to be used see Appendix 4.

The first step is therefore to decide which memory location(s) are to be used. For example, if memory location 9000 is to be addressed

?9000 means 'the contents of memory location 9000'

so the statement

10 ?9000 = 55

will make the memory location 9000 take on (i.e. store) the value 55. Note that since the indirection operator ? addresses only one memory location (1 byte) it can only store a whole (integer) number between 0–255.

Data may be retrieved from memory in a similar way, i.e.

20 D = ?9000

will make the variable D take on the value stored in memory location 9000. D would be 55 in our first example. Memory location 9000 would however still retain the value 55 for future use as could be demonstrated by the statement PRINT ?9000.

When writing programs the memory location will need to be incremented so that each new data element will be stored in a different memory location. This is most easily achieved by setting a variable equal to the first memory location to be used, i.e.

```
10  M = 9000
```

and data can be stored as follows:

```
 5  CLS
10  M = 9000
20  FOR N = 1 TO 10
30  LET B = ADVAL(1)/256
40  PRINT TAB(2,N);B
50  ?(M + N) = B
60  NEXT N
```

and retrieved in the following manner:

```
70  FOR N = 1 TO 10
80  PRINT TAB(18,N);?(M + N)
90  NEXT N
```

Line 10 sets the initial memory location, line 30 ensures that the incoming A–D converter value (0–65520) can be stored in one byte by dividing by 256. Line 50 sets memory location (M + N), i.e. 9001, 9002 etc. equal to the incoming value of B. Line 80 prints out the value stored in memory location (M + N), i.e. the values stored in memory locations 9001–9010.

The indirection operator ? addresses only one byte, as shown on p. 16. Providing that the incoming value can be reduced to the range 0–255 this provides the most efficient use of scarce memory.

Larger numbers can be stored using the indirection operator ! (pling). Pling is used in exactly the same fashion as ? but uses four memory locations at a time. Thus ! could have been used in the last program simply by changing lines 30, 50 and 80 as follows

```
 5  CLS
10  M = 9000
20  FOR N = 1 TO 10
30  LET B = ADVAL(1)
40  PRINT TAB(2,N);B
50  !(M + (N*4)) = B
```

```
60  NEXT N
70  FOR N = 1 TO 10
80  PRINT TAB(18,N);!(M+(N*4))
90  NEXT N
```

The use of ! allows storage of any number between 0 and 2^{32} but the cost is wanton use of valuable memory. Interfacing in biology rarely requires storage of numbers greater than 65536 (2^{16}) and numbers of this magnitude could be stored in only 2 memory locations (2 bytes).

If data with values in the range 0–65536 is to be saved whilst conserving valuable memory the instructions MOD and DIV can be used.

You may have realised by now that

$$65536/256 = 256$$

Any number up to 65536 can be divided by 256 and the integer answer can be stored in one byte while the remainder can be stored in another byte.

The instruction DIV divides a whole number and leaves the integer answer ignoring the remainder while the instruction MOD divides a whole number and gives the remainder, while ignoring the integer answer.

Thus a number can be split into two parts and stored separately, for example

$$42500 \text{ DIV } 256 = 166 \qquad (42500/256 = 166.015625)$$
$$42500 \text{ MOD } 256 = 4$$

The number 166 can be stored in 1 byte (the integer answer) whilst the number 4 (the remainder) can be stored in a second byte. To pull the number back out of memory, it is necessary to read the memory location containing the integer answer (= 166), multiply this by 256 (256 × 166 = 42496), and add to it the contents of the memory location containing the remainder (42496 + 4 = 42500). For example

```
  5  CLS
 10  M = 9000
 20  FOR N = 0 TO 40 STEP 2
 30  LET B = ADVAL(1)
 40  PRINT TAB(0,N/2);B
 50  ?(M+N) = B DIV 256
 60  ?(M+N+1) = B MOD 256
 70  NEXT N
 80  FOR N = 0 TO 40 STEP 2
 90  PRINT TAB(18,N/2);(?(M+N)*256) + ?(M+N+1)
100  NEXT N
```

All advantages have some cost, and the cost of conserving memory here is a slight increase in run time (5 milliseconds per sample). For programs not running at full speed this may however be a useful device to conserve memory.

The unbracketed DIM

Once the art of addressing memory locations directly has been mastered, a slightly more sophisticated approach can be used. Normally, as shown on p. 12, an array is declared by a DIM instruction in the form

    DIM C(1000)

This would allow the saving of 1000 data elements in 5000 bytes, since five bytes are reserved for each element.

However the unbracketed DIM instruction simply reserves the number of bytes indicated, i.e.

    DIM C 1000

would reserve 1000 bytes somewhere in memory.

The start address of these 1000 memory locations is stored in the variable C. To address any particular memory location use ? (query) and ' (pling) in exactly the same manner as before, i.e.

```
10  DIM S 20
20  FOR P = 1 TO 20
30  INPUT R
40  ?(S + P) = R
50  NEXT P
```

or

```
10  DIM S 20
20  FOR P = 1 TO 20 STEP 4
30  INPUT R
40  !(S + P) = R
50  NEXT P
```

If the start address is needed, insert

    15 PRINT S

The advantages of this technique are several;
(i) The initial memory location for data storage does not have to be decided, the computer does that itself:

(ii)   Because the computer chooses the start address of the data, the program will work in any mode and with subsequent program additions, providing sufficient memory is available;

(iii) If sufficient memory is not available a warning will be given early in the program rather than crashing into memory half way through.

4. Saving on Disc and Tape

One of the most useful facilities on all microcomputers is the ability to save and load information. The SAVE and LOAD commands allow storage of a particular program on disc or tape and recall into the computer's own memory at will. Equally useful however is the facility to save and load experimental data although this is achieved by different means.

There are many occasions when it will be necessary to save data, some examples of which are:
(i) To avoid losing data when the computer is switched off;
(ii) To save a complete memory bank from RAM because the computer's own memory cannot store all the data;
(iii) To build up a bank of data over an extended period;
(iv) To store data continuously rather than storing it in RAM.

Two main mechanisms are available for storing data:
(i) Saving memory banks from RAM; and
(ii) Saving variables, arrays and strings.

This chapter will deal specifically with storage on disc although disc and tape storage are identical (except where indicated). Tape storage however is extremely slow.

Before storing data either an empty formatted disc is needed, or, if the disc already contains some data, the available storage space may be optimized by entering * COMPACT into the computer (* COMPACT is not available with tape storage).

Saving memory banks from Random Access Memory (RAM)

Saving data from specific memory locations

Data which has been stored in specific memory locations using indirection operators ?, ! or $ may be saved and reloaded using the * SAVE and * LOAD commands. To achieve this the initial and final memory locations of the data must be known.

In order to use * SAVE and * LOAD commands it is essential to use hexadecimal notation to address the memory locations. This presents very few problems since the computer can be used to convert between decimal and hexadecimal notation. To obtain the hexadecimal number the instruction PRINT ~ is used. Thus PRINT ~ 12000 followed by RETURN will give the hexadecimal number. When using the BASIC language the computer will only recognize that a number is in hexadecimal if that number is prefixed by

& (known as ampersand). Consult Appendix 4 for an explanation of hexadecimal notation.

For example, imagine that some data has been stored from memory location 12000 (&2EE0) to location 15000 (&3A98). (Remember you can convert to hexadecimal notation by typing PRINT~ 12000.) This complete bank of memory can be saved using the command

    * SAVE filename   start address   finish address

    * SAVE DATA   2EE0   3A98

and may be reloaded into those memory locations by

    *LOAD DATA

The hexadecimal notation in the *SAVE command is essential since the BASIC interpreter is bypassed in this process and the machine operating system is addressed directly. This means that the BASIC language is not being used and the operating system understands hexadecimal numbers only and does not require & as a prefix. Since the BASIC language is bypassed the operating system does not require inverted commas around the filename nor does it actually discriminate between upper and lower case characters, so

    *save data   2EE0   3A98

would have performed equally well.

Problems may arise if this data bank is to be stored several times during the course of running the program since each bank of data will overwrite the last one on disc. This is due to the fact that the filename – DATA in this case – will remain unchanged.

It is possible, however, to update the filename as often as required in the following fashion.

```
   5 CLS
  10 M = 12000
  20 FOR X = 1 TO 3
  30 FOR N = 1 TO 3000
  40 LET B = ADVAL(1)/256
  50 ?(M + N) = B
  60 NEXT N
  70 ON X GOSUB 100,120,140
  80 NEXT X
  90 END
 100 *SAVE DATA1 2EE0 3A98
 110 RETURN
```

```
120  *SAVE DATA2 2EE0 3A98
130  RETURN
140  *SAVE DATA3 2EE0 3A98
150  RETURN
```

This program will save a total of 9000 bytes of data in three separate blocks. Lines 30 to 60 save data from memory location 12000 to 15000 which is saved under the filename DATA1 on disc. The process is then repeated saving the files DATA2 and then DATA3. (N.B. the RETURN in line 110 cannot be placed at the end of line 100 as :RETURN since in line 100 the operating system is being addressed directly and BASIC instructions are ignored.)

Saving screen graphics

The *SAVE and *LOAD commands have an additional useful attribute. Since screen memory is also held in RAM (both text and graphics) the current screen details can also be stored on disc.

In Chapter 3 it was shown that the screen memory locations were as follows:

MODE	Memory Location
0, 1, 2	12288–32768 (&3000–&8000)
4, 5	22528–32768 (&5800–&8000)

Thus the screen details may be saved by

| MODE 0, 1, 2 | *SAVE SCREEN 3000 8000 |
| MODE 4, 5 | *SAVE SCREEN 5800 8000 |

To reload the screen it is necessary first to place the computer into the correct MODE for that particular screen then type

    *LOAD SCREEN

screen memory will then be loaded and run automatically. (Note that there is no significance in using the filename SCREEN, any filename could have been used.)

Saving variables, arrays and strings

If data has been stored in a series of variables, in an array or as a string, the user will have no idea exactly where the computer has stored the data in memory. Although there are ways of finding this out it is cumbersome and beyond the scope of this book. Without knowing precise memory locations the *SAVE command cannot be used.

BBC BASIC contains a number of instructions which enable the saving and reloading of data in variables, arrays and strings. These are

        OPENOUT
        OPENIN (or OPENUP in BASIC II)
        PRINT #
        INPUT #
        CLOSE #

For example the statement

        10 D = OPENOUT ''FILENAME''

opens a channel (D) and allows a new file (called filename) to be written to disc. Using

        20 PRINT#D,B

data stored in the variable B will pass from the computer to disc, down channel D where it is saved. The variable (D in this case) simply holds the channel number for all communication between computer and disc. This channel must finally be closed:

        30 CLOSE#D

Thus

        10 C = 200
        20 D = OPENOUT ''DATA''
        30 PRINT#D,C
        40 CLOSE#D

would open a file called DATA and print the value of C to disc. The communicating channel is then closed.

Alternatively it may be desirable to save the contents of an existing array. This could be achieved using a program of the following type:

```
10  DIM C(20)
20  D = OPENOUT "DATA"
30  FOR N = 1 TO 20
40  PRINT#D,C(N)
50  NEXT N
60  CLOSE#D
```

This program would open a file called DATA at channel D and PRINT 20 elements (C(1) to C(20)) to disc.

It is most important that the channel is closed once the communication to disc is complete. In developing programs the user may find some frustration if the program breaks whilst the communicating channel is open to disc. If an attempt is made to rerun the program it will return with

'File open at line ____

This may be remedied by typing CLOSE#channel (CLOSE#D in this case) followed by RETURN.

Having saved data on a file the data can be re-entered at any stage as follows:

```
10  DIM C(20)
20  B = OPENIN "DATA"
30  FOR N = 1 TO 20
40  INPUT#B,C(N)
50  NEXT N
60  CLOSE#B
```

Line 20 opens a channel for input to the computer whilst line 40 inputs the 20 elements of data into the array C(1) to C(20).

The incoming data can also be manipulated between the OPENIN and CLOSE instructions. For example line 40 could be replaced with:

```
40  INPUT#B,X
42  LET C(N) = X*3
44  PRINT C(N)
```

Continuous downloading to disc

One advantage of the OPENOUT and OPENIN commands is that once a file is open data can continue to be passed to that file until the channel is closed. Since a disc can hold far more data than the computer's own RAM, this facility often overcomes the problem of memory shortage.

Data is printed (PRINT#) to a disc in a well defined format. Integer numbers are stored in five bytes, the first byte being &04 so that the computer can recognize the following four bytes as an integer number when reloading. Real numbers require six bytes of disc memory (starting with &FF) whilst strings require one byte for each character plus two additional bytes (one being &00 and the second signifying the length of the string).

As seen in Chapter 3 memory is a valuable commodity and the use of five or six bytes to store a single piece of data is excessive. Fortunately two additional instructions:

    BPUT#

and

    BGET#

allow storage and retrieval of an 8 bit (0–255) number in a single byte on disc. This is much the same as the action of the indirection operator ? on RAM.

Using these instructions a very large data base (limited only by the size of the disc storage) can be saved during the course of an experiment. For example

```
10  F = OPENOUT "DATA"
20  FOR N = 1 TO 50000
30  LET B = ADVAL(1)/256
40  BPUT#F,B
50  NEXT N
60  CLOSE#F
```

This program would open a file called DATA on channel F and take 50000 samples from A–D converter channel 1. Since this value may fall in the range 0–65520 it is divided by 256 to ensure that it can be stored in one byte and each value of B is then stored sequentially on disc.

Once the experiment was complete the data could be re-entered using

```
70   F = OPENIN "DATA"
80   FOR N = 1 TO 50000
90   S = BGET#F
100  NEXT N
110  CLOSE#F
```

Note that the BGET# command operates rather differently from INPUT#.

This program would be rather useless in practice since the variable S would be updated each time and at the end of the program S would contain only the final value of B.

This problem may be overcome using the additional instruction:

PTR#

which acts as a pointer for a particular channel (i.e. PTR#F) and points to the next byte of data to be read. When a file is opened PTR# is set to zero ready to read or write the first byte. As a byte is read from the disc file the position of PTR# is updated automatically. The pointer can be set to any position in the file to read any byte just like ?(M + N) can be used to read any byte in RAM. Thus any analysis which can be performed on data in RAM can equally well be performed on data stored on disc. Two additional instructions which are of use in this respect are:

EXT# which gives the number of bytes in the file; and

EOF# which signals the end of the file.

(NB The instructions PTR# and EXT# are not available when saving on tape.)

For example the last program could be rewritten to perform a data analysis (see Chapter 6):

```
 70  F = OPENIN "DATA"
 80  LET Y = 1
 90  REPEAT
100  PTR#F = Y: LET S = BGET#F
110  PTR#F = Y + 5: LET T = BGET#F
120  PTR#F = Y + 10: LET U = BGET#F
130  IF S < T AND T > U PRINT; "THERE IS A PEAK AT Y = ";
     Y + 5
140  LET Y = Y + 1
150  UNTIL EOF#F
160  CLOSE#F
```

Line 100 sets the pointer initially to the first byte (Y = 1), inputs that byte and places it in variable S.

Line 110 sets the pointer to the Y + 5 byte (i.e. byte 6 initially) and inputs its value into variable T. Line 120 does the same for the Y + 10 byte (byte 11 initially), the value being placed in the variable U.

Line 130 compares the values of S, T and U and continues to search for peaks. The program will end when it reaches the end of the file (line 150).

Once familiarity with the use of files is achieved, two or more files can be opened simultaneously. This can be useful when recording several different pieces of information and where each element is to be stored separately. This presents no problem providing each file is addressed by a different communicating channel (variable), i.e.

```
 10  X = OPENOUT "ONE"
 20  Y = OPENOUT "TWO"
 30  FOR N = 1 TO 10000
 40  LET B = ADVAL(1)/256
 50  LET C = ADVAL(2)/256
 60  BPUT#X,B
 70  BPUT#Y,C
 80  NEXT N
 90  CLOSE#Y
100  CLOSE#X
```

This will store 10000 values of B in file ONE (from A–D converter channel 1), and 10000 values of C in file TWO (from A–D converter channel 2).

5. Data Presentation

A biologist carrying out an experiment needs to decide how to present data – as a graph, histogram, table of results etc. Before looking at Data Analysis it is therefore useful to look at the text and graphics capabilities of the BBC Microcomputer with a view to presenting data in the most appropriate way.

Modes

There are 8 different display modes on the BBC, available by using the instructions MODE 0 to MODE 7. Modes 0, 1, 2, 4 and 5 select the resolution of the graphics, the number of colours available and the number of columns required for text characters (see the BBC User Guide). Modes 3, 6 and 7 are purely text modes. This book, for the sake of simplicity, will concentrate on the use of modes 0 and 4 since data presentation can be adequately carried out with a 2 colour monitor.

Figure 5.1 shows the format of the monitor screen in the 2 graphics modes which the authors find most useful (MODE 0 and 4). The monitor screen can be considered as a having two grid systems, one for text and one for graphics characters.

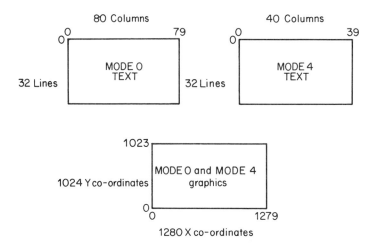

Fig. 5.1   The format of the monitor screen in MODE 0 and MODE 4 showing the grid systems for text characters and graphics.

Text

The text zero is at the top left hand side of the screen and text characters can be printed at a defined position by using the appropriate column and line numbers in a PRINT TAB(X,Y) instruction, e.g.

    PRINT TAB(0,0);"A"

will print an A at the top left hand side of the screen whereas

    PRINT TAB(39,30);"A"

will print the A at the bottom right hand corner in MODE 4 and at the bottom centre in MODE.0. If the PRINT instruction is used alone then printing will start at the top left hand side of the screen and move down one line with each PRINT instruction. The X co-ordinate value on the grid, i.e. the column number, is always given first in the TAB instruction. (Note the commonest mistake to make when using the TAB instruction is to put a space between the TAB and the brackets following: there must be no space.)

Graphics

The graphics zero is at the bottom left hand side of the screen. In both of the modes shown in Fig. 5.1 it can be seen that there are the same number of X and Y co-ordinates. The difference lies in the resolution, i.e. the size of the pixels (graphic characters). The smallest graphics characters (pixels) are only available in MODE 0. The pixels in MODE 4 are twice the size of those in MODE 0.

There is an invisible cursor for graphics which is at the bottom left hand corner of the monitor screen (grid position 0,0) when the computer is switched on. The MOVE instruction followed by graphics X,Y co-ordinates will move the invisible cursor to that X,Y position. The DRAW instruction followed by new X,Y co-ordinates will draw a line from the position of the invisible cursor to the DRAW X,Y position. Try running the program below and then vary the X,Y co-ordinates.

    10 MODE 0
    20 MOVE 150,150
    30 DRAW 1000,150

Note the same principles apply as described for the TAB instruction in that the X position (across the screen) is always specified first, but this time the co-ordinates do not have to be in brackets nor are there any problems regarding positions of spaces.

Printing a table of results

One of the simplest ways of presenting data is in the form of a table. This can be achieved using MOVE/DRAW instructions to mark out the table and PRINT TAB to correctly position the text and numbers. The program given below is very simple. Using PRINT instructions headings are printed across the top of the screen in positions defined by TAB(X,Y) co-ordinates. A loop is then set up to position the text cursor under each heading in turn to allow the values entered to be printed in the correct columns. Each circuit of the loop adds one to a counter to ensure that the next set of values are printed below the last set. Note the use of TAB with the INPUT statement. The inverted commas in lines 120 and 130 give a flashing cursor without the question mark being left on the screen.

```
 10  REM TABLE OF RESULTS
 20  MODE 4
 30  PRINT TAB(0,0);"Temperature"
 40  PRINT TAB(0,1);"Degrees Celsius"
 50  PRINT TAB(20,0);"Heart Rate"
 60  PRINT TAB(20,1);"Beats/sec"
 70  MOVE 0,930
 80  DRAW 1279,930
 90  MOVE 630,0
100  DRAW 630,1023
110  LET A = 4
120  INPUT TAB(0,A);"  "B
130  INPUT TAB(20,A);"  "C
140  LET A = A + 1
150  GOTO 120
```

Graphs

It is essential to be able to draw, scale and label the axes of a graph as well as being able to plot points and join them. However three things appear to be a problem at first:
(i)   If the origin of a graph is placed at the 0,0 co-ordinate there is no room left for labels etc.;
(ii) Text and graphic characters need to be matched up in order to label the graph;
(iii) Plotted points on a graph need to be distinguished from the lines joining those points.
All of these problems can be solved by the use of VDU instructions.

Moving the graphics zero

The VDU 29 statement followed by X,Y co-ordinates allows the position of the graphics zero to be moved, e.g.

        10 VDU 29,150;150;

This program will move the X,Y, graphics zero to position 150,150 allowing plenty of room to label and scale the axes of a graph using the X,Y co-ordinate 0,0 as its new origin. Confusion can be avoided if the VDU 29 statement is put at the start of the graph drawing program so that the same graphics zero is used throughout. Note the comma after the 29 and the semi-colons after both the X and Y co-ordinates; the program will malfunction if these are not present.

Matching text and graphics characters

This can be achieved by using the VDU 5 statement which instructs the computer to use graphics co-ordinates to position text characters instead of the text column and line numbers. Thus

```
10 MODE 4
20 VDU 5
30 MOVE 300,300
40 PRINT;"O"
50 VDU 4
```

will print the letter O at the graphics co-ordinate 300,300. The top left hand corner of the letter O is placed at the position 300,300. It is essential to put a semi-colon after the PRINT instruction when using the statement VDU 5 since this semi-colon means PRINT immediately. The statement VDU 4 returns the computer to the use of TAB or normal PRINT for specifying text character positions; if this is not done and the program is listed, the screen does not clear and continually overwrites itself. Should this happen typing VDU 4 followed by RETURN allows the screen to start scrolling again.

User defined graphics

Distinguishing plotted points on a graph from the lines joining them is achieved by surrounding the point with a user defined character, i.e. placing a character like a square or a triangle (which you, the user, have defined yourself) around the plotted point. This can be achieved by yet another VDU statement – VDU 23. This allows you to define your own graphic character.

The first number following the VDU 23 statement, (e.g. VDU 23,224) is the code used to label the particular user defined graphic. The numbers 224 to 255 are available for user defined graphics so if several lines are to be plotted on the same graph then several characters could be defined and labelled as 224, 225, 226 etc.

The principle of defining your own graphics characters is shown in Fig. 5.2. The squares in the columns across are given values 1, 2, 4, 8, 16, 32, 64, 128 and the lines are numbered from 1 to 8. The squares need to be filled in to produce the required character. Then working from line 1 downwards the value of the squares filled in are added up across the horizontal plane and placed in the VDU 23 statement separated by commas. Thus VDU 23,224,24,36,66,129,129,66,36,24 specifies the hollow diamond shape shown in Fig. 5.2. The 24 comes from 16 + 8 in line 1, the 36 from 32 + 4 in line 2, the 66 from 64 + 2 in line 3, the 129 from 128 + 1 in line 4 etc. Understanding of the principles of this method of defining your own graphics characters is enhanced by a study of bits and bytes which are covered in Chapter 3 and Appendix 3. It is essential to specify all 8 lines in addition to the VDU 23,224, since failure to do so has widespread effects.

To use a character once it has been defined is done by the statement PRINT CHR$(224); with the number being the code for the particular user defined graphic required (note the semi-colon after the brackets meaning immediately). As with the matching of text and graphic characters VDU 5 will allow a MOVE instruction preceding the PRINT CHR$(224); to

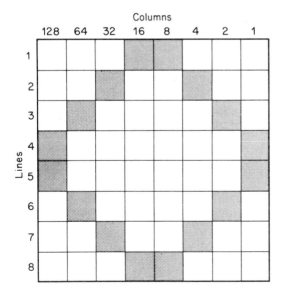

Fig. 5.2 The principle involved in defining a graphics character.

specify the plotting position on the screen. As with printing text using the VDU 5 statement, the top left hand corner of the user defined character will be positioned at the X,Y co-ordinates and small adjustments to the co-ordinates will be necessary for centering the user defined character.

The program below illustrates three different user defined graphics – a diamond, a square and a cross.

```
 10  MODE 4
 20  VDU 5
 30  VDU 23,224,24,36,66,129,129,66,36,24
 40  MOVE 400,500
 50  PRINT CHR$(224);
 60  VDU 23,225,255,129,129,129,129,129,129,255
 70  MOVE 500,500
 80  PRINT CHR$(225);
 90  VDU 23,226,129,66,36,24,24,36,66,129
100  MOVE 600,500
110  PRINT CHR$(226);
120  VDU 4
```

Note that the user defined character can be defined by the VDU 23 statement at any point in the program prior to its use.

Constructing the axes of a graph, scaling and labelling them

The first step is to decide which graphics mode to use, then the graphics zero can be redefined and the axes drawn as for example in the program below. In this section one long program is split up for ease of understanding.

```
10  REM DRAWING THE AXES OF A GRAPH
20  MODE 0
30  VDU 29,150; 150;
40  MOVE 1100,0
50  DRAW 0,0
60  DRAW 0,850
```

Thus the graphics origin has been set to 150,150 allowing room for labelling.

To scale the axes a FOR/NEXT loop is set up to move the invisible graphics cursor 100 co-ordinates along the X axis and draw a short line down each time around the loop. The minus sign in line 110 is necessary if a line is required to go below the X axis since line 30 in the above program repositioned the graphics origin (see page 32).

```
 70 REM SCALING THE AXES OF A GRAPH
 80 FOR N = 100 TO 1100 STEP 100
 90 MOVE N,0
100 DRAW N, – 10
110 NEXT N
```

A similar loop can then be used to scale the Y axis at the intervals required. Note that the X and Y axes could have been scaled in the same FOR/NEXT loop but have been kept separate for clarity.

```
120 FOR N = 100 TO 800 STEP 100
130 MOVE 0,N
140 DRAW – 10,N
150 NEXT N
```

Labelling the scale requires, in this instance, the use of the VDU 5 statement (line 170) in order to be able to match the numbers printed with the scale marks. Again a FOR/NEXT loop is used to move the graphics cursor along the scale in regular steps and the PRINT; instruction (line 210) is then implemented with each circuit. In the example below the values printed on the axes are specified by a variable A, which is increased in value with each circuit. Note again the negative values (– 25) in the MOVE statement to ensure that the labels go below the line. The N–8 in the MOVE statement ensures the centering of the number on the scale mark.

```
160 REM LABELLING THE SCALE
170 VDU 5
180 LET A = 0
190 FOR N = 0 TO 1000 STEP 100
200 MOVE N – 8, – 25
210 PRINT;A
220 LET A = A + 10
230 NEXT N
```

Labelling the Y axis can be performed in a similar way but in the following program the value of N in the FOR/NEXT loop is used both for positioning and printing the scale. Note the VDU 4 statement at the end of the labelling program to return to the TAB instruction for PRINT statements.

```
240  FOR N = 0 TO 80 STEP 10
250  MOVE −50,(N * 10) + 8
260  PRINT;N
270  NEXT N
280  VDU 4
```

Labelling the X axis is very simple since the whole label can go in a single PRINT TAB statement. Labelling the Y axis is more difficult because it may be necessary to print the title vertically down the screen. This can be done by using a FOR/NEXT loop with READ and DATA statements as shown below. One letter from the DATA statement is READ and printed each time round the loop, the value of N in the loop being used to define the TAB position.

```
290  REM LABELLING THE AXES
300  PRINT TAB(25,30);"Concentration of Potassium m.mol
     dm^ − 3"
310  FOR N = 5 TO 15
320  READ A$
330  PRINT TAB(1,N);A$
340  DATA P,o,t,e,n,t,i,a,l, ,mV
350  NEXT N
```

Plotting graphs

The simplest way of plotting data points and joining them with a line is to start with the lowest value and work through sequentially to the highest. If data is not entered in this sequence then it will be necessary to store and sort the data before plotting it.

The graph plotting program below enables the input of values of X and Y, via the keyboard, in the order in which they are required to be plotted and joined by a dotted line. Each point is surrounded by a square which is defined by the VDU 23 statement. The DRAW instruction is then used to join the plotted points with a line.

Mathematical manipulation of the values of X and Y in the MOVE and DRAW instructions is necessary in order to match the scale of the graph with the screen graphics co-ordinates, i.e. to ensure the points are plotted in the right place on the graph. In this case the manipulation is very simple (× 10) because 10 m.mol dm^{-3} takes up 100 X co-ordinates and 10 mV takes up 100 Y co-ordinates.

In the MOVE and DRAW instructions in lines 450 and 460 the values of X and Y are multiplied by 10. This is because the scale of both axes goes up in tens, i.e. 0, 10, 20, 30, etc. Since 10 scale units are equivalent to 100 X or

Y co-ordinates both inputted values need to be multiplied by 10.

The DRAW instruction produces a solid line but alternatives are available using the PLOT instruction, for example PLOT 21,X,Y would draw a dotted line. There is a large range of PLOT statements available on the BBC and it is advisable to refer to the BBC User Guide should there be additional requirements to those suggested here.

The MOVE instruction in line 450 is necessary to stop a line being drawn in response to the first DRAW instruction from wherever the invisible graphics cursor happened to be (in this case it would be at the last position of the scaling and labelling routine in line 250). The same instruction is given in line 510 to be carried out in each circuit of the FOR/NEXT loop to move the invisible cursor back to the PLOT position after printing the square around the plotted point in the position defined by the MOVE instruction in line 480. This is essential to ensure that the dotted line runs from the centre of one square to the centre of the next.

```
360  REM PLOTTING GRAPHS
370  VDU 23,225,255,129,129,129,129,129,129,255
380  PRINT TAB(30,12);"HOW MANY VALUES OF X AND Y ARE
     YOU GOING TO PLOT"
390  INPUT TAB(40,15);Z
400  PRINT TAB(30,12);"
                          ''
410  PRINT TAB(40,15);"         ''
420  FOR N = 1 TO Z*2 STEP 2
430  INPUT TAB(75,N);X
440  INPUT TAB(75,N+1);Y
450  IF N = 1 THEN MOVE X*10,Y*10
460  PLOT 21,X*10,Y*10
470  VDU 5
480  MOVE (X*10)-8,(Y*10)+16
490  PRINT CHR$(225);
500  VDU 4
510  MOVE X*10,Y*10
520  NEXT N
530  END
```

Histograms and bar charts

The first requirement is to be able to construct a bar, i.e. a filled in rectangle. This can be done by setting up a FOR/NEXT loop to draw a series of lines, of a particular length, one above the other. Lines 40 to 70 in the program below shows this construction. The MOVE instruction positions the invisible graphics cursor 400 X co-ordinates along the X axis and at one

Y co-ordinate above the X axis (in the first circuit of the loop). The DRAW instruction in line 70 then causes a line to be drawn 100 X co-ordinates long. As N increases by one each time round the loop the next line is drawn above the first and so on until the bar is 100 Y co-ordinates high.

```
10  MODE 4
20  VDU 29,200;200;
30  MOVE 1000,0:DRAW 0,0:DRAW 0,800
40  FOR N = 1 TO 100
50  MOVE 400,N
60  DRAW 500,N
70  NEXT N
```

The next step is to be able to put the bar in any required position along the X axis and to be able to increase the height of a bar. In the program below this is done by using a GOSUB routine according to the value of the number entered at line 70. Thus if a '1' is entered the GOSUB routine at line 110 is followed which draws a bar at position one on the X axis. If a '2' is entered then a different GOSUB routine at line 170 is followed to put a bar in the next position. By having the variables B1 and B2 in the GOSUB routines and adding 100 to their values each time a routine is followed, the bars can be increased in height.

In this example only 2 bars can be drawn but more subroutines could easily be added as well as more IF statements (as in lines 80 and 90) and more variables, B3, B4, B5 etc. The program would however be rather long.

```
 10  MODE 4
 20  VDU 29,200;200;
 30  MOVE 1000,0:DRAW 0,0:DRAW 0,800
 40  LET B1 = 0
 50  LET B2 = 0
 60  PRINT TAB(5,29);"Input either a '1' or a '2' "
 70  INPUT TAB(0,0);C
 80  IF C = 1 THEN GOSUB 110
 90  IF C = 2 THEN GOSUB 170
100  GOTO 70
110  FOR N = 1 TO 100
120  MOVE 0,N + B1
130  DRAW 100,N + B1
140  NEXT N
150  LET B1 = B1 + 100
160  RETURN
170  FOR N = 1 TO 100
180  MOVE 100,N + B2
```

```
190  DRAW 200,N + B2
200  NEXT N
210  LET B2 = B2 + 100
220  RETURN
```

One way of shortening the program, and in addition providing a facility to store the required data, is to use a DIM statement as shown in the program below. The entered value of C in line 60 is used on its own to position the invisible graphics cursor and draw lines in the right position along the X axis (lines 90 and 100). The same value of C is used to label the 'B' defined by the DIM statement B(C). This enables the bar to be drawn to the correct height according to the number of times that value of C has been entered (lines 70, 90 and 100). (See p. 12 for explanation of arrays.)

```
 10  MODE 4
 20  VDU 29,200;200;
 30  MOVE 1000,0:DRAW 0,0:DRAW 0,800
 40  DIM B(10)
 50  PRINT TAB(0,0);"INPUT NUMBERS FROM 0 TO 9"
 60  INPUT TAB(0,3);C
 70  LET B(C) = B(C) + 1
 80  FOR N = 0 TO 200
 90  MOVE C * 100,N * B(C)
100  DRAW 100 + (C * 100),N * B(C)
110  NEXT N
120  GOTO 60
```

Since an array of the type shown above can be used to store data, a program of the above type could be used as part of a more extensive program. The data generated and stored in the array could then be recalled by incorporating lines like 80–110 within another loop to recall all the stored data from the array.

6. Data Analysis

The preceding chapters have looked in detail at the methods available for taking data into the computer and for storing that data in memory or on disc. If this were all the computer could do it would have few significant advantages over a pen recorder or oscilloscope. One great advantage of the computer is its ability to rapidly analyse this data and present it in an appropriate form.

To appreciate the logic of data analysis within the computer it is necessary to understand fully the nature of the data which it has acquired. Perhaps the easiest way to understand this is to compare the data in the computer with the more familiar information drawn out on a chart recorder. A chart recorder provides a continuous recording of a signal. Providing the speed with which the chart paper moves is known, data analysis from the recording may simply involve the use of a ruler and an experienced eye to pick out the salient features of the waveform.

Digital data stored in the computer is different. Instead of a continuous recording the computer only stores samples of the information. This is rather like connecting equipment to a voltmeter and taking readings from the voltmeter at set intervals. Furthermore there is no specified time base to the data and, when the data is in the computer's memory an experienced eye cannot be used to pick out the required details. So before any data analysis can be performed it is essential to decide the following:

(i) How fast the incoming signal needs to be sampled in order to ensure that all its essential features are captured. If sampled too slowly some details may be missed, if sampled too quickly an excessive amount of data will be generated. There is no simple answer to this, each experiment must be considered on its merits (see Chapter 8).

(ii) How a time base for the data can be obtained. It is normally essential to know exactly when each sample was taken if accurate reproduction or analysis of data is to be achieved. Two simple procedures are shown in the following programs.

(*i*) To time a complete loop:

```
10  TIME = 0
20  FOR X = 1 TO 1000
30  LET A = ADVAL(1)
40  NEXT X
50  PRINT TIME/100;'' SECONDS''
```

(*ii*) To accurately time each sample within a slow loop:

```
10  LET T = TIME
20  FOR X = 1 TO 20
30  LET A = ADVAL(1)
40  REPEAT UNTIL TIME = T + (X * 10)
50  NEXT X
```

(iii) How the digital information is going to be analysed without using an experienced eye. This chapter is devoted to answering this question. It is not possible to cover here every conceivable analysis the user may wish to perform. Instead, an insight into possible ways of analysing typical biological signals is given. In each case the approach has been to determine the logic behind the conventional analytical approach. In other words what it is that normally enables the user to pick out the salient features and analyse a chart recorder trace. This same logic has then been applied to digital analysis on the computer.

With this approach in mind it is possible to recognize four distinct areas of analysis of a biological signal.

1. Measurement of an area.
2. Calculation of the frequency of an event.
3. Measurement of the magnitude of a response.
4. Estimation of the gradient of a line and obtaining measurements from a line.

In each case it will be necessary to decide whether this data is to be analysed 'on line' (i.e. as it comes into the computer) or whether the data is to be stored for later analysis. This will depend to a large extent on the nature of the data since the faster sampling takes place the less time is available for 'on line' analysis.

Measurement of an area

There are a number of situations where it may be required to measure the area of a particular shape. A few examples of these requirements are when it is necessary to measure the area:

(i) of a shape drawn on the visual display unit (VDU) screen using a light pen (e.g. when measuring the area of a leaf);

(ii) under a chromatographic trace (e.g. a gas-liquid or high performance chromatography trace);

(iii) under a curve on a graph (e.g. in environmental population studies).

The principle used to determine the size of an area is very simple. All that it is necessary to do is to sum together all the individual Y co-ordinate values for every X co-ordinate occupied by the shape (see Chapter 5). This approach is similar to considering the VDU screen as a piece of graph paper and the area can be calculated by adding together the number of extremely small squares on the graph paper which are enclosed by the shape. Whilst

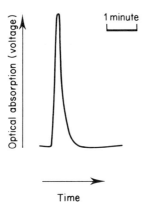

Fig. 6.1   A typical voltage output from a gas-liquid chromatogram.

this may appear to be extremely inefficient, it represents a prime case of a situation in which the computer's mathematical skills can be used to maximum benefit.

One example of an area determination is presented below. This represents the approach which might be used to analyse 'on line' the area under a chromatographic trace.

To analyse the area under a chromatographic trace (an example is illustrated in Fig. 6.1) the values entering the computer via the analogue to digital converter (the values being proportional to the voltage output of the chromatogram) are all added together.

The program below takes 1200 samples, displays them upon the screen and adds together all the samples using the variable C as a counter.

```
10  MODE 4
20  LET C = 0
30  FOR N = 1 TO 1200
40  LET B = ADVAL(1)
50  DRAW N,B/64
60  LET C = C + B
70  NEXT N
80  PRINT;"AREA UNDER CURVE = ";C
```

The value for variable C which is printed at line 80 is directly proportional to the area under the curve and is presented in arbitrary units. If necessary the program can be slowed down using a timing delay.

The major problem which is likely to be encountered with this type of analysis is that there may be a baseline level upon which the peak(s) is superimposed. This problem may be overcome by subtracting the baseline level from the value to be used to determine the area under the curve. Thus, for example, if a baseline level of 20 Y screen co-ordinates (or 1280 ADVAL 'units') was present upon a trace then line 60 in the previous program could be adapted to

60 LET $C = C + (B - 1280)$

If multiple peaks are present in a trace then it will be necessary to initially identify the peaks (see page 49) before analysis of the area under each peak can be determined.

No attempt has been made in this example to specifically calibrate the area measured. In many cases a relative area is adequate and in others it will be necessary to relate the area to that particular experiment.

Calculating the frequency of an event

It is a fairly common requirement in a biological experiment to need to measure the frequency of the occurrence of a particular event; a few examples being the measurement of heart rate from an electrocardiogram trace, drops passing through a drop counter, and animal activity from a photocell and light beam in an activity monitoring cage. Before it is possible to discuss the logic of frequency analysis it is necessary to consider the factors which may affect the users approach to the problem.

(i)   Can the computer take samples at an appropriate rate to accurately record the information?

(ii) Is the frequency the only parameter to be measured or is the magnitude of the waveform required as well?

(iii) Is the data to be analysed of an 'on'/'off' or a continuously variable nature?

(iv) Is it required to determine either the interval between every single waveform (interpulse interval), or the number of waveforms in a set period of time?

Once the decisions indicated in the questions have been made it is possible to plan a strategy for the analysis.

The problems with the rate of sampling must be the initial consideration. Generally it is likely that if a reasonable quality graphics display of the waveform is obtainable then it should be possible to analyse the data 'on line'. If however, the waveform cannot be adequately displayed on the screen, it is unlikely that it will be possible to analyse the data 'on line'. In this case one strategy which might be adopted is to store the data in memory as it enters the computer with subsequent replay of the data for analysis. If this is attempted, and it is possible to produce a reasonable graphics display from this stored data, then it will now be possible to analyse for frequency. Should the data in memory not present a true representation of the original waveform the frequency is beyond the scope of a BASIC program and the computer's 'analogue in' port. A machine code program and a fast A–D converter attached to the User port or 1 MHz bus (see Chapter 7) would therefore be needed.

In the authors' experience it is possible to analyse 'on line' up to a frequency of 1 Hz, to store in memory a frequency up to 5 Hz and with a fast A–D converter using a machine code program to analyse a signal up to 500 Hz.

The logic of frequency determinations involves the identification of a characteristic portion of that waveform. This characteristic may be either a peak, a trough, or the passage of the wave through a certain threshold value. If both the magnitude and frequency of a waveform is to be determined then it is essential to undertake peak and trough identification as described on pages 49–53, before it is possible to determine the frequency of the waveform. However if only the frequency of a waveform is needed then the passage of the waveform through a threshold value approach is probably easier. This approach involves the selection of a certain threshold value which is appropriate to the particular experiment and to detect each time the waveform passes this point. One way in which this might be achieved is illustrated below.

```
10  REPEAT
20  LET Y = ADVAL(1)
30  UNTIL Y > 18200
```

This section of program delays the program in the REPEAT–UNTIL loop until the voltage entering the analogue to digital converter in the computer exceeds a set threshold value of, in this case, 0.5 volt. (An A–D converter value of 18200 = 0.5 volt. See page 2.)

This section of program can then be combined with another REPEAT–UNTIL loop which delays the program until the A–D converter value falls below the threshold value. A combination of these two loops allows the passage of the A–D converter value through a threshold value in one direction to be detected. A program involving two such loops (lines 50 to 90 and 140 to 180) and a graphics display is presented below.

```
10  LET T1 = 0
20  MODE 4
30  *FX16,1
40  LET X = 0
50  REPEAT
60  LET Y = ADVAL(1)
70  DRAW X,Y/64
80  LET X = X + 1
```

```
 90  UNTIL Y > 18200
100  LET T2 = TIME
110  IF T1 = 0 THEN GOTO 130
120  PRINT; "INTERPULSE INTERVAL = ";(T2 − T1)/100;
     "SECONDS"
130  LET T1 = T2
140  REPEAT
150  LET Y = ADVAL(1)
160  DRAW X,Y/64
170  LET X = X + 1
180  UNTIL Y < 18200
190  IF X > 1200 THEN END
200  GOTO 50
```

This program will print out the interpulse intervals and frequency for a
waveform entering the computer's A–D converter. Lines 100, 120 and
130 are used to calculate the interpulse interval with the variable T1
holding a value for the time at which the previous waveform transversed
the threshold value of 18200. The threshold value which is set in lines 90
and 180 may be altered depending upon the magnitude and noise of the

(a)

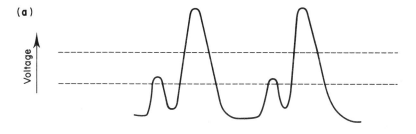

(b)

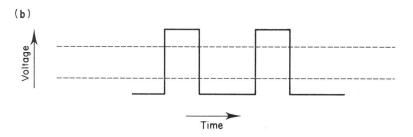

Time

Fig. 6.2   An illustration of choosing the correct threshold level for determining the
frequency of a waveform for **(a)** a continuously variable waveform, and **(b)** an on/off
waveform. The threshold is critical in **(a)** but not in **(b)**.

waveform to obtain a suitable 'trigger' level for waveform detection. If the waveform is of an on/off nature the threshold value becomes less critical as it only needs to lie between the high and low values (see Fig. 6.2).

As an alternative to determining the interpulse interval it is possible to count the number of waveforms which occur in a set time interval and hence determine the overall frequency of the waveforms. A sample program is presented below.

```
 10  LET C = 0
 20  MODE 4
 30  LET X = 0
 40  LET Z = TIME
 50  REPEAT
 60  LET Y = ADVAL(1)
 70  DRAW X,Y/64
 80  LET X = X + 1
 90  IF X > 1200 GOTO 190
100  UNTIL Y > 18200
110  LET C = C + 1
120  REPEAT
130  LET Y = ADVAL(1)
140  DRAW X,Y/64
150  LET X = X + 1
160  IF X > 1200 GOTO 190
170  UNTIL Y < 18200
180  GOTO 50
190  PRINT "FREQUENCY = ";C*60/((TIME−Z)/100);" PULSES
     PER MIN."
```

The examples illustrated in this section result in either the interpulse intervals or the frequency of the waveform being printed upon the VDU screen. However, in a completed program these values can be stored in memory for either graphical representation or further analysis later in the program.

Measuring the magnitude of a response

Many biological experiments produce an output which requires measurement of the magnitude of a given response. Figure 6.3 shows some typical traces from which a measurement of magnitude can be obtained.

As can be seen these traces divide themselves into two basic types:
(i) a progressive movement of the trace in one direction from which the

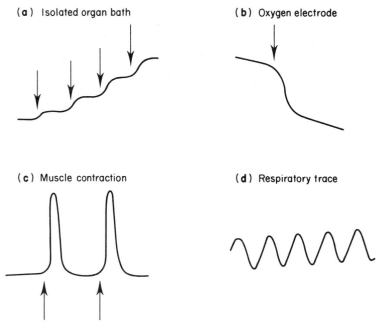

(a) Isolated organ bath

(b) Oxygen electrode

(c) Muscle contraction

(d) Respiratory trace

Fig. 6.3 Some typical biological traces from which magnitude could be measured. The arrows indicate the starting point of an event, i.e. drug addition or stimulus.

magnitude at specified intervals or of specified responses can be recorded (Fig. 6.3a, b); and
(ii) a waveform in which the magnitude from the base to the peak can be measured (Fig. 6.3c, d).

Measuring magnitude from a progressively moving trace

The types of trace illustrated in Fig. 6.3a, b represent some of the most difficult ones to try to analyse if the user has no prior knowledge of the appearance of a 'typical' trace. For example, to determine the slope of the oxygen electrode trace illustrated in Fig. 6.3b the following steps are needed:
(i)   to identify the plateau regions at the beginning and end of the trace;
(ii) to decide which portions of the trace represent the 'true' rate (i.e. excluding an initial mixing period); and
(iii) to determine the slope of this portion of the trace.
    The programming is possible but difficult. However, if the typical time period over which the changes occur is known and the appropriate section of program can be started at the commencement of the change (using an INKEY or GET instruction) the analysis can be made much simpler. In this

case the analysis may be achieved by taking a series of values at set time intervals. Such analysis can be used in any experiment where it is possible to predict that events will have occurred over previously known time intervals, for example in analysing the contraction of a piece of isolated ileum one minute after drug addition. In addition to the examples illustrated in Fig. 6.3a,b this analysis could be employed for spectrophoto-metric/colourimetric analysis, plant growth measurement, bacterial growth curves etc.

One possibility for this type of analysis is to use a slightly modified program from that illustrated in the section of data storage in memory (see page 17). A timing delay has been introduced between readings.

```
10  CLS
20  M = 9000
30  FOR N = 1 TO 10
40  LET B = ADVAL(1)/256
50  PRINT TAB(2,N);B
60  ?(M + N) = B
70  LET G = TIME
80  REPEAT UNTIL TIME = G + 1000
90  NEXT N
```

This program would take 10 readings at 10 second intervals and place the values in random access memory (RAM) locations 9001 to 9010. The intervals between readings can be adjusted by altering the delay which occurs in line 80. In this example storage of a number in a single byte using the ? (query) instruction is illustrated, but this principle could be applied to any of the other forms of memory storage described in Chapters 3 and 4.

Whilst the previous program may be satisfactory for some purposes it is likely that the user will wish to observe events which are happening in the intervals between readings using a graphical display. This may be achieved by taking rather more readings, but only storing a few of these (at set time intervals) in the computer's memory. The program below takes readings at 10 second intervals whilst maintaining a graphical display.

```
10  MODE 0
20  M = 9000
30  LET G = TIME
40  FOR N = 1 TO 1000
50  LET B = ADVAL(1)
60  DRAW N,B/64
70  IF N MOD 100 = 0 THEN ?(M + (N/100)) = B/256
80  REPEAT UNTIL TIME = G + (N * 10)
90  NEXT N
```

The majority of the program is similar to that shown previously, however line 70 allows for the selection of every hundredth value, in which case the reading is stored in a memory location. A total of 10 values will be stored in memory locations 9001–9010. Please note the use of the TIME instruction in lines 30 and 80 which is designed to reduce errors which can occur due to the time taken for the program to be executed (see page 41).

Measuring magnitude from a waveform

Many biological signals have a waveform nature (Fig. 6.3c, d) i.e. electro-cardiogram, heart muscle contraction, gas chromatogram, atomic absorption spectrophotometer, etc. The magnitude of each response is normally assessed by picking out the peak and the trough of each wave-form and measuring between the two. But how can a peak or a trough be recognized? The peak would probably be recognized by first picking out the rising phase and falling phase and then assessing the precise point at which one becomes converted into the other.

How can this be done with digital information in the computer? Assume that 1000 samples of the waveform shown in Fig. 6.4 have been taken and each sample has been stored sequentially in memory from memory location 10000 to memory location 11000 using a program of the following type:

```
10  MODE 0                50  LET A = ADVAL(1)
20  *FX16,1               60  ?(M + N) = A/256
30  M = 10000             70  DRAW N,A/64
40  FOR N = 0 TO 1000     80  NEXT N
```

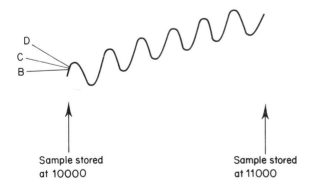

Fig. 6.4 A visual impression of one batch of data stored in memory between memory locations 10000 and 11000. B, C and D are used to sample the data in memory and are gradually moved through the memory locations during the analysis.

The stored samples now need to be gone through to find the peaks and troughs. Initially it is necessary to recognize whether the trace is going up or going down. The easiest way of doing this is to ask the computer to compare three sequential samples. This can be done by bringing samples out of memory and placing them in 3 different variables (B,C and D) i.e.

```
 90  X = 1
100  B = ?(M + X)
110  C = ?(M + X + 1)
120  D = ?(M + X + 2)
130  IF B < C AND C < D THEN PRINT "GOING UP"
140  IF B > C AND C > D THEN PRINT "GOING DOWN"
150  X = X + 1
160  IF X = 1000 THEN END
170  GOTO 100
```

X in line 90 acts as a counter to increment the memory locations to be sampled until the 1000th sample is reached in line 160. Variables B, C and D contain the digital number stored in three consecutive memory locations. If the trace was going up at the time of the samples then B (the first sample) should be less than C, (the second) and D (the third). Thus the conditional statement in line 130 would be fulfilled and the fact that the trace is on a rising phase would be confirmed. Alternatively if the trace was moving down the condition in line 140 would be fulfilled. If the trace was flat or irregular neither condition would be fulfilled and the program would simply move through to line 150 and increase the counter (X) by one. Line 170 repeats the cycle and the program would go back to line 100 and sample the next 3 memory locations.

Once it has been confirmed that the trace is definitely going up or down it is possible to start looking for a peak or a trough.

A peak may be recognized by virtue of the fact that a rising trace is suddenly converted into a falling trace. Thus once a rising trace has been recognized in line 130 the program would go into a peak searching routine (lines 300 to 360) by modifying line 130:

```
130  IF B < C AND C < D THEN GOTO 300
300  REPEAT
310  X = X + 1
320  B = ?(M + X)
330  C = ?(M + X + 1)
340  D = ?(M + X + 2)
350  UNTIL B < C AND C > D
360  GOTO 100
```

Lines 310–340 continue to sample sequential memory locations on the rising waveform. Lines 300 and 350 repeat the operation of gradually moving through the stored data in memory until the point at which the value of C is greater than the previous and subsequent memory locations is found. C therefore represents the peak.

To confirm that the trace is moving up or down the program returns to line 100 and recognizes subsequent peaks or troughs in a similar manner (see Fig. 6.5).

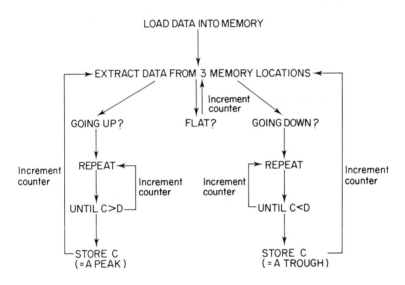

Fig. 6.5 A flow diagram illustrating the technique described in the text for finding peaks and troughs in a waveform.

In theory this program is fine but in practice it has two flaws:

(i) Normal biological signals leaving a transducer are not smooth but contain some noise. An exaggeration of this is shown in Fig. 6.6. The current program would pick out every minute component of this noisy waveform. There are several ways to overcome this problem.

   (a) Average several data points as they enter the computer and store this average value.

   (b) Sample more slowly so as to ignore minor flaws in the signal (providing you won't miss important elements of the signal).

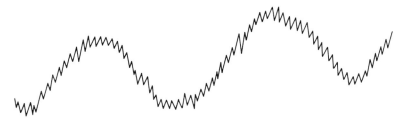

Fig. 6.6 This figure illustrates, in an exaggerated form, the noise that may be inherent in a biological signal, though not observed on a VDU screen display.

(c) When analysing stored data, instead of sampling memory locations next to each other, it may be better to spread out the samples i.e. lines 100–120 and 320–340 could become:

B = ?(M + X)
C = ?(M + X + 4)
D = ?(M + X + 8)

(ii) The second problem is that the statement in line 350 can be too precise and would require an extremely sharp peak. This is best overcome by replacing line 350 with:

UNTIL C > D

A complete program to pick out peaks and troughs is shown below. This will require modification for the user's own purposes bearing in mind the sampling rate and the nature of the signal. Once the peak and trough values have been found and stored a simple subtraction will obviously give the magnitude. A further program applied to a spirometer trace is shown in Chapter 8.

```
 10 CLS
 20 *FX16,1
 30 MODE4
 40 M = 9000
 50 FOR P = 1 TO 1000
 60 LET A = ADVAL(1)/256
 70 ?(M + P) = A
 80 DRAW P,A*4
 90 NEXT P
100 LET X = 1
```

```
110 LET B = ?(M + X)
120 LET C = ?(M + X + 4)
130 LET D = ?(M + X + 8)
140 IF X > 1000 GOTO 400
150 IF B > C AND C > D GOTO 300
160 IF B < C AND C < D GOTO 200
170 LET X = X + 1
180 IF X > 1000 GOTO 400
190 GOTO 110
200 REPEAT
210 LET B = ?(M + X)
220 LET C = ?(M + X + 4)
230 LET D = ?(M + X + 8)
240 LET X = X + 1
250 IF X > 1000 GOTO 400
260 UNTIL C > D
270 MOVE X,C*4
280 DRAW X,(C*4) + 100
290 GOTO 110
300 REPEAT
310 LET B = ?(M + X)
320 LET C = ?(M + X + 4)
330 LET D = ?(M + X + 8)
340 LET X = X + 1
350 IF X > 1000 GOTO 400
360 UNTIL C < D
370 MOVE X,C*4
380 DRAW X,(C*4) - 100
390 GOTO 110
400 END
```

Estimating the gradient of a line and obtaining measurements from a line

Gradient of a line

Biologists frequently need to assess the gradient of a line, for example when following the output from an oxygen electrode to calculate the rate of oxygen consumption – or production.

The experienced eye may be tempted to 'guesstimate' the gradient of the line but the best approach would be a statistical analysis of the data. This is a job at which the computer excels. Statistical formulae may be obtained from any suitable text (i.e. Parker, R. (1979). *Introductory Statistics for Biology*, second edition. Edward Arnold, London) and are easily converted for use in a program by breaking the problem down into its component parts.

For example, to calculate the gradient of the line use a linear regression analysis, the equation for a straight line being of the type:

$$Y = A + B*X$$

where Y and X give the relative values for any point
B is the gradient of the line
A is the point at which the regression line crosses the Y axis

The statistical equations for the gradient (B) and intercept (A) are:

$$B = \frac{\text{sum X * Y} - \dfrac{(\text{sum X * sum Y})}{n}}{\text{sum X}^2 - \left(\dfrac{(\text{sum X})^2}{n}\right)}$$

$$A = \left(\frac{\text{sum Y}}{n}\right) - \left(B * \frac{\text{sum X}}{n}\right)$$

In addition the user may assess how well the points fit the calculated line by obtaining the correlation coefficient (R) given by the equation:

$$R = \frac{\text{sum X * Y} - \left(\dfrac{\text{sum X * sum Y}}{n}\right)}{\sqrt{\left(\text{sum X}^2 - \dfrac{(\text{sum X})^2}{n}\right) * \left(\text{sum Y}^2 - \dfrac{(\text{sum Y})^2}{n}\right)}}$$

This looks horrifying in terms of programming but, in practise, from looking at the equations there are only five variables to be calculated, namely:

(a) sum of all X values (sum X) – given the variable name SX;
(b) sum of all Y values (sum Y) – given the variable name SY;
(c) sum of the product X * Y (sum X * Y) – given the variable name SXY;
(d) sum of the square of X (sum X^2) – given the variable name SXSQ;
(e) sum of the square of Y (sum Y^2) – given the variable name SYSQ;

and of course the number of samples taken for the analysis (n) will need to be known.

Each of the equations to calculate B, A or R can be solved once these variables have been calculated.

The calculation of the five variables can be performed at any point in the program once the individual X and Y values have been obtained, for example:

```
10  FOR N = 1 TO 20
20  INPUT X
30  INPUT Y
40  SX = SX + X
50  SY = SY + Y
60  SXY = SXY + (X*Y)
70  SXSQ = SXSQ + (X^2)
80  SYSQ = SYSQ + (Y^2)
90  NEXT N
```

Lines 40, 50, 60, 70 and 80 are used to obtain the sum of the required variable. Once the five variables have been calculated, taking into account all the values of X and Y, the slope B, the intercept A and the correlation coefficient R may be obtained. For example line 100 could calculate the slope (B) by:

```
100  B = (SXY − ((SX*SY)/N))/(SXSQ − ((SX^2)/N))
```

The value of A could be calculated from:

```
110  A = (SY/N) − (B*(SX/N))
```

Now any value of Y can be calculated from a known value of X using the equation:

$$Y = A + BX$$

```
120  Y = A + (B*X)
```

The line representing the best fit for the points can also be drawn knowing the relationship between X and Y values and the co-ordinates you have used to plot them originally. For example assume that the X and Y values each fall within the range 0–900 and that their raw values have been used as the X and Y graphics co-ordinates to plot the points. The line of best fit may be drawn by:

```
130  MOVE 1,(A + (B*1))
140  DRAW 900,(A + (B*900))
```

A complete example of this approach is given in Chapter 8 but you may like to try the following program:

```
10  MODE 4
20  VDU 23,224,24,36,66,129,129,66,36,24
30  FOR N = 1 TO 10
```

```
 40  INPUT X
 50  INPUT Y
 60  MOVE X − 8,Y + 16
 70  VDU5
 80  PRINT CHR$(224)
 90  VDU4
100  MOVE X,Y
110  SX = SX + X
120  SY = SY + Y
130  SXY = SXY + (X * Y)
140  SXSQ = SXSQ + (X^2)
150  SYSQ = SYSQ + (Y^2)
160  NEXT N
170  N = N − 1
180  B = (SXY − ((SX * SY)/N))/(SXSQ − ((SX^2)/N))
190  A = (SY/N) − (B * (SX/N))
200  MOVE 1,(A + (B * 1))
210  DRAW 900,(A + (B * 900))
```

Note that line 170 is necessary since on leaving the loop the value of N is 11 and not 10 as expected.

Obtaining values from a calibration graph

Many calibration graphs do not present an ideal straight line relationship. This would be true, for example, of a sodium calibration curve performed on a flame photometer (Fig. 6.7).

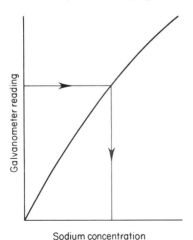

Fig. 6.7   A calibration curve for sodium determination on a flame photometer. An unknown may easily be read from the calibration curve as shown.

It is undoubtedly possible to obtain an equation which describes this line and calculate the unknown from that equation. It would normally be easier, however, to draw a line across from the Y axis until it met the curve and then draw a vertical line down from this point to find its intercept on the X axis (Fig. 6.7).

This can be achieved on the BBC microcomputer by using the instruction POINT. The purpose of this instruction is to examine the colour of a pixel at specified co-ordinates on the screen. This is achieved by a colour number being returned to POINT when particular co-ordinates have been examined. (See section on colour graphics in BBC User Guide.) Thus in a two colour mode (MODE 0) the number returned to POINT will be 0 if the pixel is black (unlit) and 1 if the pixel is white (lit). The specified co-ordinates of POINT from the value of the reading (Y value on the graph) can be moved horizontally until a lit pixel (i.e. the curve itself, when POINT = 1) is found. A vertical line to the X axis will provide the unknown value. (Further details of the use of POINT for other modes and colours can be found in the BBC User Guide.)

The following program gives an example of the use of POINT. The important section illustrating the use of POINT can be found in lines 130–170.

```
 10  MODE 0
 20  VDU 29,200;200;
 30  MOVE 800,0:DRAW0,0:DRAW 0,800
 40  MOVE 0,0
 50  FOR C = 1 TO 5
 60  INPUT TAB(2,2)"ENTER X VALUE";X
 70  INPUT TAB(2,4)"ENTER Y VALUE";Y
 80  DRAW X,Y
 90  NEXT C
100  INPUT TAB(2,6)"ENTER UNKNOWN";U
110  MOVE 0,U
120  T = 1
130  REPEAT
140  T = T + 1
150  MOVE T – 2,U
160  DRAW T – 2,U
170  UNTIL POINT(T,U) = 1
180  DRAW T,0
190  PRINT "UNKNOWN = ";T
200  GOTO 100
```

NOTE: This program will only work on MODE 0. The accuracy of this technique depends upon the resolution of the graphics. On low resolution graphics the value of T in line 120 will need to be increased.

Lines 10 to 90 simply set up a calibration graph from the X and Y values entered in lines 60 and 70.

The variable T is used as a counter to move across the X axis. The colour of co-ordinates T, U are found in line 170 and if the pixel is lit, POINT = 1. To prevent errors the value of T must never = O since the Y axis has been drawn along this line (i.e. all the pixels are lit). In addition, it is necessary to ensure that the horizontal line produced by lines 150 and 160 is always behind (T − 2) the POINT being examined.

A full program showing the use of POINT in an interfacing situation is given in Chapter 8.

7. Additional Signal Input and Output

The 'analogue in' port on the BBC microcomputer provides the simplest means of interfacing biological equipment and is the most appropriate interfacing link for many biological purposes. However it is possible that the user's requirements cannot be satisfied by the 'analogue in' port. A few situations in which this may occur are:

(i) taking information into the computer at a higher sampling rate than is available via the inbuilt A–D converter, for example, if an electro-cardiogram is to be accurately recorded 'on line';

(ii) taking in signals of an on/off nature as might be received from a switch, for example in a behavioural study to detect whether an animal passes in front of a photocell;

(iii) switching devices on or off such as peristaltic pumps, stepper motors, valves and solenoids; and

(iv) sending out a variable voltage using a digital to analogue converter.

Whilst it is not possible to cover all of these and other areas in this limited text it is possible to give an introduction to the mechanisms available for such interfacing.

In addition to the 'analogue in' port converter there are three other communication systems which may all be found on the underside of the BBC microcomputer. They are called the User port, Printer port and 1 MHz bus. Of these three it is likely that it may be desirable to connect a printer to the Printer port. Thus although it is possible to design a system around the Printer port it is probably an unwise long term strategy. The User port is relatively simpler to understand so this port is considered before the 1 MHz bus. The chapter continues with a description of some input and output devices relevant to the biologist.

The User port

The User port is an 8 data line input/output port. Each data line operates in a digital manner (i.e. 0 or 1) which represents an on/off signal. In addition to the 8 data lines there are 2 control lines, 2 low power 5V lines and 8 ground (0V) lines making a total of 20 lines to which you may connect.

Configuration of the User port and its connection to equipment

Connection of equipment to the User port can be made using a 20-way insulation displacement socket (IDC)(RS Components Ltd. Cat. No. 467–289) and a 20-way cable (RS Components Ltd. Cat. No 360–122).

The socket and lead can be assembled in a press such as supplied by Radio-spares (RS Components Ltd. Cat. No. 468–197 and 468–232). It is conventional for pin 1 of the plug (the pronged connector in the micro-computer) to be at the top right hand side of the plug when looking at the pin and the red line on the 20-way cable is usually connected to pin 1.

The arrangement of the connections to the User port are shown in Fig. 7.1 and the connections to the 20-way cable are listed in Table 7.1.

Table 7.1

Wire number (counting from the right hand side, with red wire connected to pin 1)	Connection
1	+5V
2	Control line 1 (CB1)
3	+5V
4	Control line 2 (CB2)
5	0V
6	Data line 0 (PB0)
7	0V
8	Data line 1 (PB1)
9	0V
10	Data line 2 (PB2)
11	0V
12	Data line 3 (PB3)
13	0V
14	Data line 4 (PB4)
15	0V
16	Data line 5 (PB5)
17	0V
18	Data line 6 (PB6)
19	0V
20	Data line 7 (PB7)

Connections on the User port.

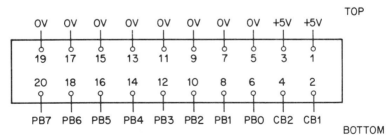

Fig. 7.1   The arrangement of the connections to the User port when viewed from the outside of the non-inverted microcomputer.

Operation of the User port

There are two principal pieces of information which are essential to an understanding of the operation of the eight data lines.

(i) Each data line can operate as an input or an output line.

(ii) Each data line can only operate as either an 'on' or 'off' device. For example if a signal output, relative to a 0V line is required, the user would be able to measure approximately 5V on the data line when it is switched 'on' and 0V on the line when it is switched 'off'. Conversely application of a 0V or a 5V signal to the data line would represent 'off' and 'on' signals respectively when using the data line for input.

The User port is controlled by a particular integrated circuit chip in the microcomputer which is called a 6522 Versatile Interface Adaptor (VIA) integrated circuit. The control of the 6522 VIA, and hence the User port, is brought about by reading from or writing to a particular area of memory called 'SHEILA'. This is achieved using the ? (query) indirection operator described in Chapter 3. SHEILA covers a range of memory locations (65024–65279) of which only two are of interest in this text. These two memory locations are 65122 and 65120 (or &FE62 and &FE60 in hexadecimal notation) and they control the User port data direction register and User port input/output register respectively. The data direction register is used to instruct the microcomputer that the data lines are being used for either input or output, whereas the input/output register is used to send or receive data to or from the User port. The system is arranged such that data line 0 (see Table 7.1) is controlled by bit 0 in memory locations 65122 and 65120, data line 1 is controlled by bit 1, data line 2 is controlled by bit 2, etc. (see Chapter 3 and Appendix 3 for an explanation of bits and bytes).

Programming the User port

Use of the data direction register

Before it is possible to communicate with equipment via the User port it is necessary to select the direction in which the information is required to travel. Generally, all the data lines will need to be either input or output lines and this may be easily achieved with one of the two following lines of program.

        ?65122 = 0

(i.e. all bits in the data direction register = 0)

will set all the User port data lines as input lines, or

        ?65122 = 255

(i.e. all bits in the data direction register = 1)

will set all the User port data lines as output lines.

It is also possible to set some data lines as output and some as input lines by entering a '0' in the appropriate bit for an input line and '1' in the bit for an output line. (See the section in Chapter 3 and Appendix 3 on bits to see the logic involved.) Thus, for example

?65122 = 15

will set data lines 0 to 3 as output lines and 4 to 7 as input lines. The bit pattern for this is

Bit   7 6 5 4 3 2 1 0
      0 0 0 0 1 1 1 1

(bits 0 to 3 in the data direction register = 1, bits 4 to 7 = 0).

Once the data direction register has been set for input and/or output on the various data lines, it will remain in that conformation until reset by changing the value in ?65122. Hence it is not necessary to set the data direction register before each input/output operation.

Use of the input/output port for output
Once the data direction register has been set, it is possible to communicate via the User port in the chosen direction. Initially output via the User port will be considered.

To switch 'on' a data line, a '1' should be entered in the appropriate bit in the User port input/output register. The data line may be switched 'off' by entering a '0' in the same input/output port register bit.

Thus to switch on and off data line 0 the following lines of program need to be used:

10 ?65122 = 255
20 ?65120 = 1

Line 10 sets the data direction register for all data lines to be output lines. Line 20 will then switch 'on' the data line.

30 ?65120 = 0

Line 30 will then switch 'off' the data line.

If it is necessary to use more than one output data line this can be easily achieved. If, for example, data lines 0 and 1 are required the following lines of program could be used.

20 ?65120 = 1

This will switch on data line 0 but switch off data line 1.

30 ?65120 = 2

Line 30 will switch on data line 1 but switch off data line 0.

    40 ?65120 = 3

Line 40 will switch on both data lines and

    50 ?65120 = 0

will switch off both data lines.

If more data lines are required they can similarly be switched on or off by placing a '1' or a '0' in the appropriate bit in memory location 65120. The number which needs to be ? (queried) into memory location 65120 may be calculated using the program in Appendix 3, or the number required for data lines 0 to 3 is shown in Table 7.2.

Such output control may easily be included into a program so that a device such as a peristaltic pump, stepper motor or light can be switched on for a period of time and then switched off again. For example the program below would switch on a device connected to data line 0 for 5 seconds. The program could be included in a larger program as a sub-routine or defined procedure which could be called up when it was necessary to switch on the device. The length of time the device is switched on could be increased or decreased by altering the delay in line 1030 (see for example the program on the pH stat in Chapter 8).

```
1000  ?65122 = 255
1010  ?65120 = 1
1020  LET B = TIME
1030  REPEAT UNTIL TIME = B + 500
1040  ?65120 = 0
```

Table 7.2

Number	Data line			
	0	1	2	3
0	off	off	off	off
1	on	off	off	off
2	off	on	off	off
3	on	on	off	off
4	off	off	on	off
5	on	off	on	off
6	off	on	on	off
7	on	on	on	off
8	off	off	off	on
9	on	off	off	on
10	off	on	off	on

Table 7.2   *contd*

| Number | Data line | | | |
	0	1	2	3
11	on	on	off	on
12	off	off	on	on
13	on	off	on	on
14	off	on	on	on
15	on	on	on	on

An illustration of the required number which needs to be ? (queried) into memory location 65120 (the User port input/output register) for switching on or off data lines 0 to 3.

Use of the input/output port for input

To use the User port as an input port it is initially necessary to set the selected data lines as input lines in the data direction register (see page 61). Information may now be fed into the data lines from an external source. When the value in memory location 65120 is read, each bit is set by the voltage applied to the line. If 0V is applied to the line a '0' will be present in the relevant bit and if 5V is applied a '1' will be present in the bit. Care must be taken to ensure that more than 5V is never applied to any of the data lines as this may cause damage to the microcomputer.

It may be convenient to adopt data lines 0–3 as output lines and 4–7 as input lines to avoid confusion when using both input and output facilities. Thus to study an input into data line 7 the data direction register is set

    10 ?65122 = 0

Then the value in the memory location can be read by

    20 LET B = ?65120

The value returned into the variable B may, for example, be the number 255. To decide if there is a '0' or '1' in bit 7 of the memory location the value in the variable 'B' should be divided by 128 using the DIV instruction which can only return a whole number as the answer. (If the number is greater than 128 then a '1' must be present in bit 7, see Appendix 3.)

    30 LET B7 = B DIV 128
    40 PRINT "BIT 7 = "";B7

If the value of the variable B7 = 1 then there is a '1' in bit 7.

It is possible to study several input data lines. In the following example, three data lines are studied. As before the data direction register is set and the value in the memory location is read

```
10  ?65122 = 0
```

(or ?65122 = 31 if data lines 0–4 are to be output lines)

```
20  LET C = ?65120
```

The values in each bit can now be determined using the following type of program.

```
30  B7 = C DIV 128
40  B6 = (C MOD 128) DIV 64
50  B5 = (C MOD 64) DIV 32
60  PRINT "BIT 7 =",B7
70  PRINT "BIT 6 =",B6
80  PRINT "BIT 5 =",B5
```

The values in the variables B7, B6 and B5 represent the figure present in the appropriate bit of memory location ?65120. Similar programming can be used for any number of the bits in this memory location. (If you are unsure of the logic of this program try replacing lines 10 and 20 by INPUT C.)

The use of osbyte calls for communication with the User port

The BBC microcomputer User Guide suggests that it is good practice not to directly address specific memory locations as has been advocated in this section. The major reason for this suggestion relates to the possibility that a second processor may be being used or may be used in the future. If a second processor is used many of the memory locations are changed including those involved with the User port and 1 MHz bus. If programs involving the direct addressing of memory locations have been written they will need to be rewritten if the user then starts to use a second processor. The alternative approach is to use the approved method of addressing the User port via an OSBYTE call, which is slightly harder than the method described in this section but is better working practice. The use of such OSBYTE calls is described in Appendix 2.

The 1 MHz bus

The 1 MHz bus is similar to the User port in that it is an 8 data line input/output port. It has, however, much greater potential than the User port as an input/output port as it is possible to connect up 254 different devices to the one set of 8 data lines. Sadly, as might be expected from such a versatile linkage, the 1 MHz bus is difficult to use.

The easiest application of the 1 MHz bus is to use it as an 8 data line input. To do this it is necessary to connect the 8 data lines shown in Fig. 7.2 to the input device via a 34-way IDC socket and ribbon cable in a similar manner to that used for the User port. The programming of the 1 MHz bus does not require the setting of a data direction register as with the User port. To take a reading from the 1 MHz bus the following line of program is required:

    10 LET C = ?64704

(or LET C = ?&FCC0 in hexadecimal notation)

This sets the variable C to the 8 bit reading from the 8 data lines on the 1 MHz bus.

Any further uses as either a single or multichannel output device, or as a multichannel input device, are beyond the scope of this publication and you are advised to consult other texts. Two problems associated with these uses are given below.

(i)   Unlike the User port output data lines, those of the 1 MHz bus are not latched. This means that when an output is sent to the data lines the voltage change only lasts a few microseconds, whereas on the User port the voltage change is held until the data lines are reset.

(ii) If more than 8 input or output data lines are to be used it will be necessary to use the address lines (pins 27–34 in Fig. 7.2). The address lines will need to be decoded by a multiplexing device. There may also be problems in decoding the address lines and 'clean up' circuits may be required.

Input devices

Probably the two most likely devices to be used as input devices to the

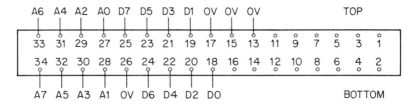

Fig. 7.2   The arrangement of the connections to the 1 MHz bus when viewed from the outside of the non-inverted microcomputer. The data lines (D0–D7) are indicated together with several earth lines (0V) and the address lines (A0–A7).

User port are analogue to digital converters and switch inputs. The uses of these devices are described below.

Analogue to digital converters

As there are already four analogue to digital converter channels on the rear of the BBC microcomputer it may appear rather excessive to be considering additional ones. However it is possible that in certain applications four A–D converters are insufficient or it may be found that the maximum rate of sampling (approximately once per 10 milliseconds) is not fast enough for the application(s) to be studied (i.e. if nerve action potentials are to be recorded).

Before considering the use of such devices it is necessary to be aware that although an A–D converter can achieve a high rate of conversions, if the program is in BASIC even the simplest of programs is incapable of taking readings at less than 5 millisecond intervals. To overcome this problem three possible solutions are given below.

(i) Purchase fast A–D converters with associated software to run the converter. Phillip Harris is one supplier of such devices and of software suitable for such applications. This approach is fine if the manufacturers' software suits the user's requirements exactly.

(ii) Purchase or construct A–D converters which can be programmed by the user. For example Palmer Bioscience supply a fast A–D converter which can be programmed by the operator in either BASIC or assembly language (full instructions for BASIC programming of the unit are supplied with the unit). With this unit samples can be taken at 20 microsecond intervals if the program has been written in assembly language.

Alternatively Radiospares Components Ltd, offer a range of A–D converters, which can be assembled into a complete device from circuit diagrams available in their data sheets. One such suitable A–D converter integrated circuit chip is the RS ZN448. However please be warned that a considerable degree of electronic knowledge is necessary before construction is commenced.

This approach allows a maximum degree of flexibility for the programmer; but due to the relative slowness of BASIC programs it will be necessary either to learn how to write programs in assembly language (not something to be taken lightly) or to use a machine code compiler such as that produced by ACK Data. A machine code compiler is a program which can convert a program written in BASIC to one written in machine code.

(iii) Use a data logging device such as the 'VELA' unit supplied by Data Harvest. This allows for easy programming and flexibility of usage for the programmer as the VELA is an independent data store, but may be a relatively expensive provision if such a facility is not often required.

Switch inputs

A switch input is a simple device which allows the computer to detect if something is on or off. One example of the use of a switch input might be in a behavioural experiment where an animal presses a lever. If a switch input is required this can be easily achieved using a circuit similar to the one illustrated in Fig. 7.3.

If the switch is in the 'off' position 5 volts will be applied to the line and a '1' will be present in the appropriate bit of memory location 65120 (see page 64) when the User port is read; a '0' will be present if the switch is in the 'on' position.

The switch may be replaced by a light activated switch circuit (such as the RS Components Ltd. Cat. No. 305–434, incorporated into the circuit outlined in RS Data Sheet 2107) which will then allow the microcomputer to detect low and high light levels.

Output devices

There are a range of different output devices which are of use to biologists. In our opinion the three most useful ones are switching relays, digital to analogue converters (D–A converters) and stepper motors. The uses of these devices are described below.

Switching relays

Switching relays are probably the most useful of all output devices as they enable electrical devices to be switched on or off. It is possible to think of many things requiring switching as part of a program to control an experiment. A few examples are provided below:

(i)   Controlling a peristaltic pump to make additions of known amounts of liquid (see the pH stat program in Chapter 8);

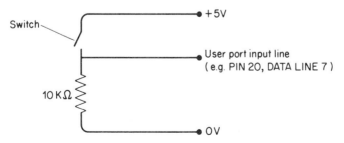

Fig. 7.3 A switch input circuit.

(ii)   Switching on and off the chart drive of a pen recorder to record only selected parts of an experiment;

(iii) Control of lighting and heating in a plant growth chamber; and

(iv)   Control of a physiological/pharmacological stimulator to produce a complex pattern of stimulus parameters.

Although all switching relays perform the same task there are two types required to cover the full range of possible applications. One type is a relay designed to switch mains voltages safely and the other a low voltage relay with a variety of applications including the switching of equipment remote control lines.

Griffin & George sell an excellent mains switching relay (Cat. No. CRB-060-010W) which is a self contained unit (the Cat. No. for the lead is ECA-260-W). It can be plugged into the mains supply and has a switched mains output to a mains socket. This mains switching relay has an input for a 5 volt switching input which is compatible with the output from the User port (labelled TTL input). To connect the TTL input to the mains switching relay it is necessary to produce a 20-way IDC lead as described on page 59 and separate wires 5 (0 volts) and 6 (data line 0) from the rest of the cable. Solder a 2 mm black banana plug to wire 5 and a red banana plug to wire 6. The remaining wires should be carefully insulated from each other and from making any contact with any other electrical equipment. They can either be left as long as the wires to the mains switching unit; or, for a neater cable, they can be separated from wires 5 and 6 back to the IDC socket and then cut off neatly and insulated at the socket. The red and black banana plugs can now be plugged into the corresponding coloured sockets on the mains switching relay. The device can be switched on and off by switching data line 0 (see page 62). Similarly, any other data line can be used for this purpose.

It is often considered to be good electrical practice to place a so called 'buffer' between the computer User port output and the connections to external equipment to protect the computer from damage due to mishandling and/or equipment malfunction. In an attempt to maintain this publication at a relatively simple level buffers have not been included in the suggested scheme for wiring up equipment to the User port. It is therefore essential that any connections made using leads such as the one suggested for the mains switching relay, are connected *before* any of the equipment involved is switched on. It is also very important to check all connections carefully before use and of course observe the correct polarity for all connections. It is possible to 'buffer' the computer's output by using a Darlington Octal Driver (R.S. Components Ltd Cat. No. 303–422) or a Griffin & George Octal Buffer Board (Cat. No. CRB–030–A) as shown in Fig. 7.4.

If a low voltage switching relay is required to switch a device such as a physiological/pharmacological stimulator by remote control or to apply power to a low voltage device then the Griffin & George mains switching relay is unsuitable for this purpose. It will be necessary either to purchase a

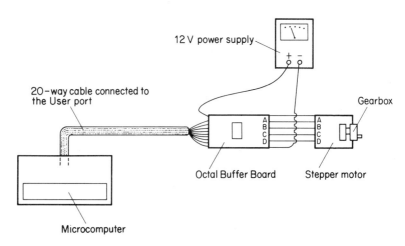

Fig. 7.4   A circuit diagram for an Octal Buffer Board and 12V relay system.

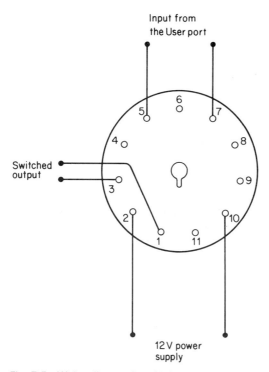

Fig. 7.5   Wiring diagram for a high-sensitivity transistorized relay.

suitable low power switching unit (the Unilab and Interbeeb units have suitable switching relays incorporated) or to build your own.

If building your own unit a suitable relay is the R.S. Components Ltd, high sensitivity transistorized relay (Cat. No. 348–245) which is very simple to wire up and only requires a 12 volt laboratory power supply to operate it. A circuit diagram is presented in Fig. 7.5. Alternatively, by using an Octal Buffer Board as described in Fig. 7.4, it is possible to use a conventional relay, such as R.S. Components Ltd, Cat. No. 346–946, in place of a high-sensitivity transistorized relay. The relay can be switched on and off by switching on and off data line 0 (see page 62). Alternatively this relay or the mains switching relay may be connected to any of the other data lines (by connecting to a 0V line and the appropriate wire at the User port for the chosen data line). In this case the relay will be switched by switching on or off output on the data line which has been selected.

Thus for a switching relay connected to the data line '0' after the data direction register has been set by

?65122 = 255

(or ?65122 = 1 for only data line '0' as an output line)

The relay can be switched on by

?65120 = 1

and the relay can be switched off by

?65120 = 0

Digital to analogue converters (D–A converters)

Digital to analogue converters, as the name implies, have the opposite effect of A–D converters, i.e. they take digital signals from the computer and convert them into an analogue or variable voltage. Thus it is possible for the computer to output a variable voltage using a D–A converter.

If a D–A converter is connected to the User port then it needs to be connected to all 8 data lines on the User port. The User port is then set so that all data lines are output lines, i.e.

?65122 = 255

The voltage output from the D–A Converter will then vary in response to the output from the computer via the User port data lines. If all the data lines are switched 'off' then the output from the D–A converter will be 0 volts, i.e.

?65120 = 0

will give a 0 volt output.

If all the data lines are 'on' the output from the D–A converter will be at the maximum level (usually 2.5 or 5 volts), i.e.

    ?65120 = 255

will give the maximum voltage output.

If some of the computer's data lines are 'off' and some 'on' an inbetween voltage will be obtained. This may be predicted from the following formula:

$$V = \frac{A \times M}{255}$$

where   V = D to A converter output voltage
         A = the number in the range 0–255, ? (queried) into memory
              location 65120
         M = maximum output voltage of the D–A converter.

Thus to output half the maximum voltage of the D–A converter the following line of program would be required:

    ?65120 = 127

It is possible to include such instructions in a longer program. For example, to output a variable voltage from the maximum D–A converter output voltage down to 0 volts in 255 steps with each step taking one second, the following program would perform this task:

```
10 ?65122 = 255
20 FOR N = 255 TO 1 STEP − 1
30 ?65120 = N
40 LET B$ = INKEY$(100)
50 PRINT N
60 NEXT N
```

Depending upon the user's approach to computing the use of a D–A converter may be considered unnecessary as it is often possible to achieve a similar end result using just a computer and a printer. Two examples of uses for a D–A converter are given below.

(i) Where it is required to produce a continuous output to another recording device (such as a pen recorder) of values that have been manipulated by the computer. For example if recording a heart beat using the computer accompanied by a continuous recording of heart rate on a pen recorder. The computer could analyse the information and output a voltage proportional to heart rate via the User port and a D–A converter to a pen recorder.

(ii) If the computer is to be used as a data store (or transient recorder). It is possible to record the particular event (such as an action potential) and store the data in the computer. The data can then be replayed via a D–A converter onto a slow response recorder.

There are several commercially available digital to analogue converters all of which are parts of particular manufacturers' systems (e.g. Unilab, Philip Harris and Griffin & George all market one as part of their system). Alternatively, it is possible to construct your own D–A converter unit. For example, R.S. Components Ltd, sell a range of suitable integrated circuit chips (e.g. ZN425E, ZN435, DACO800, ZN428E) which may be used. Full information on the construction of such devices is available in the R.S. Components Ltd. data sheets.

Stepper motors

Stepper motors are electric motors which can be controlled by a computer. The name 'stepper motor' is derived from the fact that the motor can be made to rotate in a series of equal sized steps (e.g. steps of 7.5 degrees). Thus if an instrument requires a rotary motion it can be controlled by a

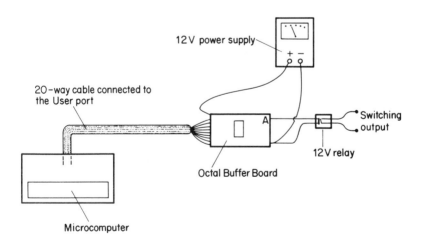

Fig. 7.6 The arrangement of a typical stepper motor system. The connections between the User port and the Octal Buffer Board are such that Data line 0 (pin 6) on the User port is connected to terminal A on the Buffer Board, Data line 1 (pin 8) is connected to terminal B, Data line 2 (pin 10) is connected to terminal C, etc.

computer. There are numerous examples of situations where a rotary motion might be required. A few possibilities are: to control instrument zero or gain controls; to alter the wavelength on a spectrophotometer; and to obtain a linear motion by rotating the spindle on a rack and pinion.

One supplier of stepper motors is Griffin & George who sell a suitable system for use in scientific environments. Unfortunately it is not possible to just purchase a stepper motor unit and run it directly from the computer, it is also essential to purchase associated equipment. The required equipment is an Octal Buffer Board, a 12 volt power supply, a gearbox and a User port connection lead. The arrangement of a typical system is shown in Fig. 7.6.

The gearbox in the system is to allow the motor to be geared down to increase the amount of torque which is developed. This is necessary as the torque development of small stepper motors is generally insufficient to turn an instrument spindle.

The stepper motor may be made to rotate clockwise by using the following program:

```
 10  ?65122 = 255
 20  ?65120 = 9
 30  LET Z = INKEY(20)
 40  ?65120 = 5
 50  LET Z = INKEY(20)
 60  ?65120 = 6
 70  LET Z = INKEY(20)
 80  ?65120 = 10
 90  LET Z = INKEY(20)
100  GOTO20
```

For counterclockwise rotation of the stepper motor the following program will be required.

```
 10  ?65122 = 255
 20  ?65120 = 10
 30  LET Z = INKEY(20)
 40  ?65120 = 6
 50  LET Z = INKEY(20)
 60  ?65120 = 5
 70  LET Z = INKEY(20)
 80  ?65120 = 9
 90  LET Z = INKEY(20)
100  GOTO20
```

8. Sample Problems and Programs

The following programs have been chosen to illustrate in a more detailed way many of the principles put forward in the earlier chapters of this book and to bring various elements together into a complete working program.

Problems of interfacing a bell spirometer

The conventional Benedict-Roth bell spirometer (Fig. 8.1) is commonly used to measure the basal metabolic rate (BMR) of a subject at rest. The principle of operation is that the subject breathes pure oxygen from the bell. As the subject breathes out, carbon dioxide is absorbed in the side arm. Thus the volume of the bell is reduced by an amount corresponding to the volume of oxygen consumed by the subject. Over a period the volume of the bell falls and provides a convenient measurement of oxygen consumption. The kymograph trace would look like that in Fig. 8.2.

Drawing the line of best fit through the peaks or troughs in the waveform will give the overall slope of the line and thus oxygen consumption can be measured over a set period.

Interfacing the spirometer to the microcomputer, to perform this experiment automatically, introduces several problems.

Firstly the spirometer, like many items of biological equipment, provides

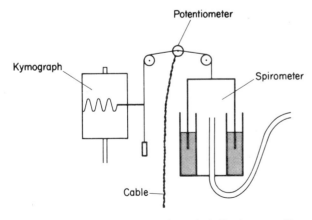

Fig. 8.1 A part cross-sectional view of a bell spirometer. Note the potentiometer spindle around which the cord is wrapped.

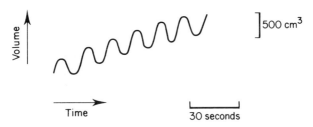

Fig. 8.2   The output from a bell spirometer when measuring basal metabolic rate.

no electrical output. Nevertheless it is normally possible to provide an output by some alternative means. In this case the writing pen could have been attached to an extended arm on an isotonic transducer. A simpler approach however is to secure a 10 turn 10 K potentiometer next to the cord connecting the bell to the writing pen. The cord is then wrapped once round the potentiometer shaft so that movement of the bell causes rotation of the potentiometer.

The wiring of the potentiometer is shown in Fig. 8.3. The output from the potentiometer may be fed into the 'analogue in' port via a signal conditioning unit or, by adding the additional resistor and variable potentiometer shown in Fig. 8.3, the voltage may be adjusted to 1.8V and fed directly to the 'analogue in' port.

Secondly, having obtained an electrical signal it is necessary to decide how to program the computer. This can be broken down into several stages.

(i) A calibration of the spirometer onto the computer will be required. This can be obtained by setting the spirometer to a known value, i.e. 5 litres, and entering the voltage through the A–D converter (using ADVAL). The spirometer can be set to a new volume, i.e. zero litres, and that voltage entered. The difference between the two inputted A–D values thus equals the difference in spirometer volumes.

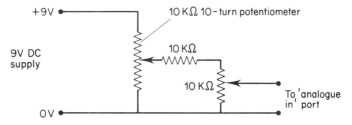

Fig. 8.3   Circuit diagram for direct input into the 'analogue in' port from a potentiometer.

(ii) Having calibrated the system samples need to be taken over a convenient period. To analyse the input it is necessary to pick out all the peaks and all the troughs and then find the line of best fit. The sampling time needs to be sufficiently fast to ensure that the true peaks or troughs in the waveform are not missed. However, if the samples are taken excessively quickly RAM will soon fill up and the experimental period will be reduced. Respiratory rate may be expected to be in the range of 12 to 20 per minute, thus a peak and trough will appear approximately every 3 seconds. To ensure accurate detection it is necessary to take at least 100 samples over this time, i.e. a sample every 30 milliseconds. The sampling time can be controlled by the TIME statement.

(iii) Sampling at this rate allows very little time for 'on line' analysis so the incoming data needs saving in RAM or on disc (or both). Once the data has been saved it can be analysed by the peaks and troughs routine covered in the data analysis section (Chapter 6).

(iv) To perform a linear regression on all the peaks or troughs two co-ordinates for each peak or trough value are used – namely its Y (volume) value and the X (time) value at which it was recorded. The timing can be obtained from the original sampling loop (using the I value of that loop) and, knowing the memory location of each peak or trough value, the original time of that sample can be found. These two co-ordinates will need storing in memory, as they are recognized, for later regression analysis (as covered in data analysis).

(v) A linear regression analysis will give the line of best fit through the peaks or troughs. The original calibration will then allow two separate calculations of oxygen consumption, one from line through peaks and one obtained from troughs. These can be compared, or the average of these two calculations can be obtained.

(vi) Once this stored data has been analysed there are a number of choices:
    (a) terminate the experiment;
    (b) save all the data for the future;
    (c) repeat the experiment.

The following program would perform this analysis.

```
10  REM INITIALIZE VARIABLES
20  ML = 0
30  USED = 0
40  VOL1 = 0:VOL2 = 0
50  CAL1 = 0:CAL2 = 0
60  Z = 0:W = 0
70  DU = 0
80  CLS
90  *FX16,1
```

```
100  REM SPIROMETER CALIBRATION
110  PRINT TAB(6,9);"SET SPIROMETER TO ZERO LITRES"
120  PRINT TAB(8,11);"THEN PRESS RETURN"
130  X = GET
140  IF X = 13 THEN LET CAL1 = ADVAL(1)/256
150  PRINT TAB(8,13);"ZERO CALIBRATION = ";CAL1
160  PRINT TAB(4,17);"NOW SET SPIROMETER TO 5 LITRES"
170  PRINT TAB(8,19);"THEN PRESS RETURN"
180  Y = GET
190  IF Y = 13 LET CAL2 = ADVAL(1)/256
200  PRINT TAB(8,21);"5 LITRE CALIBRATION  = ";CAL2
210  TIME  = 0: REPEAT UNTIL TIME = 200
220  CLS
230  PRINT TAB(5,4);"CONNECT SUBJECT TO SPIROMETER"
240  PRINT TAB(5,6);"COMMENCE EXPERIMENT"
250  PRINT TAB(5,9);"AFTER ONE MINUTE PRESS RETURN"
260  PRINT TAB(5,11);"AND CONTINUE RECORDING "
270  PRINT TAB(5,13);"FOR TWO MINUTES"
280  S = GET:IF S = 13 GOTO 300
290  GOTO280
300  CLS
310  REM   SET MODE AND
320  REM   INITIAL MEMORY LOCATIONS
330  MODE 4
340  P = 17000
350  T = 18000
360  M = 11000
370  REM   TAKE IN SAMPLES
380  REM   STORE AND TIME
390  REM   DU TIMES THE LOOP
400  TIME = 0
410  FOR H = 1 TO 5000
420  LET A = ADVAL(1)/256
430  ?(M + H) = A
440  DRAW H/5,A*4
450  Y = INKEY(2)
460  NEXT H
470  DU = TIME
480  REM   SEARCH THROUGH MEMORY
490  REM   UNTIL TRACE GOING UP
500  REM   OR GOING DOWN
510  LET H = 1
520  LET B  = ?(M + H + 3)
530  LET C = ?(M + H + 5)
540  LET D = ?(M + H + 7)
550  IF H > 5000 GOTO 940
```

```
560  IF B > C AND C > D GOTO 780
570  IF B < C AND C < D GOTO 620
580  LET H = (H + 1)
590  IF H > 5000 GOTO 940
600  GOTO 520
610  REM   FINDING PEAKS
620  REPEAT
630  LET C = ?(M + H + 3)
640  LET D = ?(M + H + 5)
650  LET H = H + 1
660  IF H > 5000 GOTO 940
670  UNTIL C > D
680  REM   STORE PEAK VALUE
690  REM   AND TIME (H VALUE)
700  ?P = C
710  ?(P + 1) = H/20
720  P = P + 2
730  Z = Z + 1
740  MOVE H/5,C*4
750  DRAW H/5,(C*4) + 100
760  GOTO 520
770  REM   FINDING TROUGHS
780  REPEAT
790  LET C = ?(M + H + 3)
800  LET D = ?(M + H + 5)
810  LET H = H + 1
820  IF H > 5000 GOTO 940
830  UNTIL C < D
840  REM   STORE TROUGH VALUE
850  REM   AND TIME (H VALUE)
860  ?T = C
870  ?(T + 1) = H/20
880  T = T + 2
890  W = W + 1
900  MOVE H/5,C*4
910  DRAW H/5,(C*4) - 100
920  GOTO 520
930  REM   LINEAR REGRESSION ON PEAKS
940  P = 17000
950  LET SX = 0:LET SY = 0:LET SXY = 0:LET SXSQ = 0 :LET
     SYSQ = 0
960  FOR N = 1 TO (Z*2) STEP 2
970  Y = ?((P + N) - 1)
980  X = (?(P + N))*4
990  SX = SX + X
1000 SY = SY + Y
```

```
1010  SXY = SXY + (X*Y)
1020  SXSQ = SXSQ + (X^2)
1030  SYSQ = SYSQ + (Y^2)
1040  NEXT N
1050  LET N = (N – 1)/2
1060  LET B = (SXY – ((SX*SY)/N))/(SXSQ – ((SX*SX)/N)
1070  LET A = (SY/N) – (B*(SX/N))
1080  MOVE 1,(A + B)*4
1090  DRAW 1000,(A + (B*1000))*4
1100  VOL1 = 1000*B
1110  REM    LINEAR REGRESSION ON TROUGHS
1120  T = 18000
1130  LET SX = 0:LET SY = 0:LET SXY = 0:LET SXSQ = 0 :LET
      SYSQ = 0
1140  FOR N = 1TO (W*2) STEP 2
1150  Y = ?((T + N) – 1)
1160  X = (?(T + N))*4
1170  SX = SX + X
1180  SY = SY + Y
1190  SXY = SXY + (X*Y)
1200  SXSQ = SXSQ + (X^2)
1210  SYSQ = SYSQ + (Y^2)
1220  NEXT N
1230  LET N = (N – 1)/2
1240  LET B = (SXY – ((SX*SY)/N))/(SXSQ – ((SX*SX)/N))
1250  LET A = (SY/N) – (B*(SX/N))
1260  MOVE 1,(A + B)*4
1270  DRAW 1000,(A + (B*1000))*4
1280  TIME = 0:REPEAT UNTIL TIME = 300
1290  CLS
1300  VOL2 = 1000*B
1310  REM    CALCULATION OF
1320  REM    OXYGEN CONSUMPTION
1330  ML = 5000/(CAL2 – CAL1)
1340  PRINT TAB(6,2);"DURATION OF EXPERIMENT = ";
      (INT(100*DU/6000))/100;"MINS"
1350  PRINT TAB(6,7);"RATE OF OXYGEN CONSUMPTION"
1360  PRINT TAB(6,9);"(PEAKS) = ";INT((VOL1*ML)/
      (DU/6000));"ml/min"
1370  PRINT TAB(6,14);"RATE OF OXYGEN CONSUMPTION"
1380  PRINT TAB(6,16);"(TROUGHS) = ";INT((VOL2*ML)/
      (DU/6000)); "ml/min"
1390  PRINT TAB(6,21);"MEAN OXYGEN CONSUMPTION = ";
      INT((((VOL2*ML)/(DU/6000)) + ((VOL1*ML)/
      (DU/6000)))/2);"ml/min"
```

```
1400  PRINT TAB(11,30); "PRESS Y TO RUN AGAIN"
      TAB(18,31)"N TO END"
1410  F = GET:IF F = 89 GOTO 330
1420  END
```

A description of the program is presented below.

Lines 20–70
initialize the following variables:
> CAL 1 and CAL 2 store the 0 and 5 litre ADVAL calibration values;
> ML, USED, VOL 1 and VOL 2 are variables used in the final calculation;
> DU times the initial sampling loop so that the timing of each sample can be obtained;
> Z and W are used to store the number of peaks and troughs respectively.

Lines 110–200
set up the initial calibration and use delays (lines 130 and 180) together with prompts for the operator.

Lines 340–360
set up the memory locations for storing the initial data (M) and after analysis the peaks (P) and troughs (T).

Lines 400–470
form the initial sampling loop, which stores and draws the incoming data and times (DU) the overall sampling. Line 450 controls the sampling rate.

Lines 510–920
search through the stored data in memory and pick out the peaks and troughs as covered in Chapter 7. Once the peaks and troughs are found their value is stored in memory together with the H value of the existing loop. This H value provides the means of retrieving the time for this peak or trough from the overall (DU) timing.

Lines 940–1300
perform a linear regression analysis on the peaks and troughs (see Chapter 7). This analysis uses as its X and Y co-ordinates the peak (or trough) value and the stored H value. The H value stored fell within the range 0–250 (H/20 in lines 710 and 870). For the regression this value is multiplied by 4 in lines 980 and 1160 so that it falls in the range 0–1000.
The regression lines are calculated and drawn from the A and B values (lines 1060–1090 and lines 1240–1270).

Lines 1330–1390

perform the final calculation. VOL 1 and VOL 2 are used to calculate (in uncalibrated units) the amount of oxygen used. This is obtained from the equation for the line.

NOTE $(A + B*1000) - (A + B*0) = 1000*B$

The variable ML is used to store the digital number corresponding to a volume of one ml. from the original calibration.

Lines 1360 and 1380 calculate the final oxygen consumption from the equation $\dfrac{\text{Volume in ml}}{\text{time in minutes}}$.

Lines 1400–1410

give the option of terminating or repeating the experiment. An option of storing the existing data to disc using a *SAVE instruction could also have been included here.

A program to run a pH stat

In some biological situations it is necessary to measure acid secretion or production, for example when measuring either gastric acid secretion or activity of an acid producing enzyme.

If the acid producing enzyme is considered, it would be possible to investigate pH change as a measure of enzyme activity. A faster rate of pH change would reflect a greater enzyme activity. This approach has two problems, one being the logarithmic nature of the pH scale, and the other that the activity of the enzyme may be pH dependent. Both of these problems can be overcome if a pH stat is used, in this case the pH is constantly monitored and alkali is added whenever the pH falls below a set value. Thus when using a pH stat it is rather like undertaking a continuous acid-base titration.

To set up a pH stat an input and an output system is required in addition to the microcomputer and its related equipment. The input system is composed of a pH meter and electrode, and a signal conditioning unit. The output system is composed of a mains switching relay and a peristaltic pump. The arrangement of the system is shown in Fig. 8.4.

Initially it was necessary to check the voltage output from the pH meter to determine how it related to the pH. The majority of pH meters appear to have a voltage output which increases with increasing pH. There is a wide range of pH meter voltage outputs and in the majority of cases it will be necessary to condition the voltage before input into the computer. One pH meter was found to have an output which ranged from $-0.7V$ (pH 0) to 0V (pH 7) to $+0.7V$ (pH 14) in which case a 'back off' unit would be needed to adjust the voltage output. A pH was selected at which the pH was to be maintained (pH 7 in this case) and using a simple program similar to the one

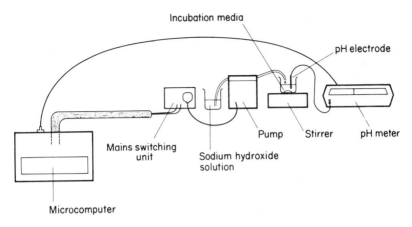

Fig. 8.4   The arrangement of the equipment for a computerized pH stat.

listed below, the signal conditioning unit was adjusted to give a set A–D converter value (30000) at the selected pH of 7.

```
10  A = ADVAL(1)
20  PRINT A
30  S = INKEY(50)
40  GOTO 10
```

For the output system it was decided to run the peristaltic pump for a set period of time (one second). The concentration of the sodium hydroxide solution and speed at which it was delivered were set manually. For the sake of convenience it was decided that the pump should deliver 100 microlitres of solution in one second and for the particular experiment envisaged this required 0.25 micromoles of sodium hydroxide (i.e. a 2.5 mmol dm^{-3} sodium hydroxide solution).

The program is written in two sections, the initial section allows the pH to be adjusted to the set level before any measurements are taken. The second section of the program takes measurements of the amount of alkali added and builds up the bar chart display over five minute periods.

The program is designed around feedback loops, a similar loop existing in both sections of the program. Each loop commences by taking an A–D converter reading. This value is compared to a set value (30000) that represents a certain pH (in this case pH 7). The comparison is made using the IF instruction and the following question is asked, 'Is the A–D converter reading above or below the set value?'. If the answer to the question is 'above', then the set pH has been reached or slightly exceeded. In this case if the program is in the first section it proceeds to the second section. If the program is in the second section no action is taken and the program returns to the beginning of the loop. If the answer to the question is 'below', then the pH is below the set value and alkali needs to be added. This is achieved in a subroutine which runs the peristaltic pump for one second and then waits for 10 seconds to allow for mixing of the solution to occur. The program then returns to the commencement of the loop.

```
 10  REM * * * INITIAL ADJUSTMENT TO THE SET pH * * *
 20  MODE7
 30  LET K = 1
 40  ?65122 = 255
 50  VDU 23;8202;0;0;0;0;
 60  PRINT TAB(2,3);"THE pH STAT IS SETTING ITS SELF UP"
 70  REPEAT
 80  LET B = ADVAL(1)
 90  IF B < 30000 THEN GOSUB 510
100  LET D = TIME
110  REPEAT UNTIL TIME = D + 250
120  UNTIL B > 30000
130  LET K = 0
140  MODE4
150  VDU 23;8202;0;0;0;0;
160  REM * * * DRAWING THE GRAPH * * *
170  PRINT TAB(10,3);"pH STAT"
180  MOVE 1160,200
190  DRAW 200,200
200  DRAW 200,840
210  FOR N = 200 TO 840 STEP 64
220  MOVE 200,N
230  DRAW 190,N
240  NEXT N
250  FOR N = 200 TO 1160 STEP 96
260  MOVE N,200
270  DRAW N,190
280  NEXT N
290  FOR N = 0 TO 10 STEP 2
300  PRINT TAB(4,(25 - N * 2));N
310  NEXT N
320  FOR N = 12 TO 15
330  READ TITLES$
340  PRINT TAB(0,N);TITLES$
350  NEXT N
360  DATA H + ,ADDED,      ,(umol),
370  FOR N = 5 TO 50 STEP 5
380  PRINT TAB(4 + N * 3/5,27);N
390  NEXT N
400  PRINT TAB(15,29);"TIME (MINUTES)"
```

```
410  REM * * * COLLECTION OF DATA OVER 5 MINUTES * * *
420  FOR N = 1 TO 10
430  MOVE 104 + (96 * N),200
440  LET A = TIME
450  REPEAT
460  LET B = ADVAL(1)
470  IF B < 30000 THEN GOSUB 510
480  UNTIL TIME > A + 29999
490  NEXT N
500  END
510  REM * * * SUBROUTINE * * *
520  LET Z = TIME
530  ?65120 = 1
540  REPEAT UNTIL TIME = Z + 100
550  ?65120 = 0
560  IF K = 1 GOTO 630
570  FOR F = 1 TO 8
580  PLOT 0,0,1
590  PLOT 1,96,0
600  PLOT 0,0,1
610  PLOT 1, - 96,0
620  NEXT F
630  REPEAT UNTIL TIME = Z + 750
640  RETURN
```

A description of the program is presented below.

Lines 10–60
Line 30 sets a 'flag' which is used later in the program.
Line 40 sets the data direction register for the User port.
Line 50 switches off the flashing cursor.
Line 60 prints the initial screen display.

Lines 70–130
A 'REPEAT UNTIL' loop is set up between lines 70 and 120. The condition
being waited for in line 120 is that the variable B is greater than a set value
(in this case 30000). The value for variable B is obtained in line 80 from the
A–D converter which is detecting the pH via the pH meter. In the event that
the value of B is less than 30000 and hence the pH is below the set value
the computer is instructed, in line 90, to go to a subroutine. The subroutine
is the part of the program which controls the peristaltic pump. Therefore
starting at line 80 a value is taken in via the A–D converter. At line 90 the
question is asked, 'Is the value above or below the set value ?'. If the
answer to this question is 'above' the program continues and will be
continued through past line 120. If the answer is 'below' the program goes
to the subroutine which adds alkali and waits a short time. Upon return

from the subroutine the program is sent back to line 70 and then the question is asked again. This sequence will be repeated until the value of B is greater than 30000.

Line 130
This line resets the 'flag' set in line 30 (the flag is used in the subroutine).

Lines 140–400
This sequence of lines draws, scales and labels a bar chart.

Lines 410–490
These lines set up a loop which is used to draw the bars on the bar chart. Line 430 moves the graphics cursor to the base of each bar on the bar chart. Lines 450–480 operate in a similar manner to lines 70–90 and 120, however on this occasion instead of repeating until a certain value is reached the sequence is repeated until a period of 30000 centiseconds (5 minutes) has elapsed. As before if the value is below the set value (30000) then the program jumps to the subroutine which adds alkali.

Lines 510–640
Lines 520–550 switch on the pump for 1 second.
Line 550 refers to the 'flag' set in either line 30 or line 130. If K = 1 then the program is in the initial display and jumps to line 630. If K = 0 then the bar chart is displayed and line 570 to 620 are performed next.
Lines 570–620 draw in a section of the bar chart. This is done using relative MOVE(PLOT 0,X,Y) and DRAW(PLOT 1,X,Y) instructions. The graphics cursor moves up one line and draws a line across, then moves up a line and draws a line back, this is repeated 8 times using a FOR-NEXT loop. Thus once the bar chart has been drawn a section of the 'bar' is inserted each time an addition of alkali is made.
Line 630 times a delay to allow for mixing of the added alkali and line 640 returns the program from the subroutine.

The program may be adapted for other pHs, or for situations in which alkali is being produced.

To alter the 'set' pH either the amplification of the signal conditioning unit could be adjusted or the value in lines 90, 120 and 470 could be altered. An increased value in lines 90, 120 and 470 would represent a higher pH when using the majority of pH meters.

To alter the system such that it may be used when alkali is added the following changes are necessary. The peristaltic pump should be filled with an appropriate acid solution and the 'less than' sign (<) in lines 90, 120 and 470 should be changed to a 'greater than' sign (>).

It is also possible to alter the range over which the pH stat operates. If, for example, it was required to increase the amount of alkali added then the scale for the Y axis could be altered in line 300 and the title for the scale altered in line 360.

'On line' spectrophotometric determinations

One of the commonest experiments in biology is finding the concentration of a particular substance (protein, carbohydrate, lipid, ion) in a solution.

A typical experiment might involve making up a range of standard concentrations of the substance being investigated – obtaining readings from for instance, a spectrophotometer, of absorbance for these standards and using these readings to plot a calibration curve or line of best fit. Readings can then be obtained from the unknown solution and the calibration curve can be used to find the concentration of the unknown.

The equipment involved may be a spectrophotometer, colourimeter, flame photometer etc. Most of this equipment has an electrical output and if there is no obvious output it is often easy to obtain one. An output from an old flame photometer has been obtained by connecting leads to the input of its sodium and potassium meters.

The output can be connected to the 'analogue in' port on the back of the BBC microcomputer via a signal conditioning unit to bring the signal from the equipment to within the 0–1.8 volt range of the computer's A–D converter. It is necessary to measure the output of the equipment and to see how that output varies from minimum to maximum since while most apparatus of this type works on a 0 volt output for a zero reading and, for example + 5 volts for maximum readings the opposite can be the case for certain equipment (i.e. + 5V for a zero reading).

Having obtained an electrical output which can be brought to within the required range it is necessary to decide how to program the computer. This can be broken down into separate steps.

(i) A setting up routine whereby the output from the equipment can be matched with the program is needed, i.e. it is necessary to calibrate the equipment. In this case this is achieved by adjusting the variable control on the signal conditioning unit to bring the maximum output from the equipment to a certain preset level.

(ii) After this setting up procedure the graph can be drawn, scaled and labelled.

(iii) The next step is to find out how many standards are being used since this can typically vary from only 2 (including a blank as zero) to as many as 10. This is done using a printed question and an input instruction.

(iv) A sequence of prompts by the computer is the next stage to tell the experimenter when to enter the concentrations of the standards and when to 'press the space bar' in order to get the computer to accept the reading (output) from the equipment. The points can then be joined or a line of best fit constructed.

(v) The last stage is to obtain readings for the unknown samples and again this is done by the computer giving prompts. Once the reading is taken the concentration of the unknown can be calculated. This could be done mathematically from the line of best fit or, as in this example, POINT can be used (see Chapter 6). The POINT instruction allows the user to read a value for the unknown just as if a calibration curve had been constructed on a piece of graph paper. The reading from the unknown can be marked off on the Y axis and thus it can be found where a line drawn horizontally cuts the calibration curve. Dropping a line vertically downwards from this point gives the concentration from where this line cuts the X axis. The computer can do all of this using the POINT instruction so that these lines can be seen being drawn across and down with the subsequent printing of the concentrations.

```
 10   REM INTERFACING A SPECTROPHOTOMETER
 20   REM CALIBRATION CURVE FOR GLUCOSE
 30   REM CALCULATING UNKNOWN GLUCOSE
      CONCENTRATIONS
 40   REM SETTING UP ROUTINE
 50   MODE 0
 60   PRINT TAB(0,4);"Adjust the gain control on the signal
      conditioning unit until number below = 800 when
      spectrophotometer gauge reads full scale deflection"
 70   PRINT TAB(0,30);"Press space bar when adjustment is
      complete"
 80   LET B = ADVAL(1)/65520*800
 90   PRINT TAB(38,15);B
100   TIME = 0
110   REPEAT UNTIL TIME = 50
120   C$ = INKEY$(0)
130   IF C$ = "   " GOTO 150
140   GOTO 80
150   REM DRAWING THE AXES OF A GRAPH
160   L = 2
170   MODE 0
180   VDU 29,250;200;
190   MOVE 1129,0: DRAW 0,0: DRAW 0,873
200   REM SCALING THE AXES
210   FOR N = 100 TO 1000 STEP 100
220   MOVE N,0: DRAW N, – 10
230   NEXT N
240   FOR N = 80 TO 800 STEP 80
250   MOVE 0,N: DRAW – 10,N
260   NEXT N
270   REM LABELLING THE SCALE
280   VDU 5
290   A = 0
300   FOR N = 0 TO 1000 STEP 100
310   MOVE N – 16, – 16: PRINT;A
```

```
320  A = A + 1
330  NEXT N
340  A = 0
350  FOR N = 0 TO 800 STEP 80
360  MOVE - 80,N + 8:PRINT;A
370  A = A + 0.1
380  NEXT N
390  VDU 4
400  REM LABELLING THE SCALE
410  PRINT TAB(25,28);''Concentration of Glucose mmol
     dm - 3''
420  FOR N = 1 TO 10
430  READ A$
440  PRINT TAB(8,N);A$
450  DATA A,b,s,o,r,b,a,n,c,e
460  NEXT N
470  REM PLOTTING GRAPHS
480  PRINT TAB(25,12);''HOW MANY VALUES OF THE
     STANDARDS ARE YOU GOING TO PLOT''
490  PRINT TAB(25,13);''INCLUDE THE BLANK IN THIS NUMBER
     (as a zero standard)''
500  INPUT TAB(40,15);Z
510  PRINT TAB(25,12);''
                        ''
520  PRINT TAB(25,13);''
                        ''
530  PRINT TAB(40,15);''      ''
540  FOR N = 1 TO Z
550  PRINT TAB(0,30);''INPUT THE STANDARD
     CONCENTRATION''
560  INPUT TAB(0,28);X
570  PRINT TAB(0,30);''PLACE STANDARD IN
     SPECTROPHOTOMETER AND PRESS SPACE BAR TO
     REGISTER THE READING''
580  C$ = INKEY$(0)
590  IF C$ = ''  ''GOTO 610
600  GOTO 580
610  LET Y = ADVAL(1)/65520*800
620  IF N = 1 THEN MOVE X*100,Y
630  DRAW X*100,Y
640  PRINT TAB(0,30);''
                                ''
650  NEXT N
660  PRINT TAB(0,28);''      ''
670  PRINT TAB(0,29);''      ''
680  REM FINDING CONCENTRATION OF UNKNOWN
690  PRINT TAB(0,30);''PLACE UNKNOWN IN
     SPECTROPHOTOMETER AND PRESS SPACE BAR TO
     REGISTER THE READING''
```

```
700  PRINT TAB(0,29);''          ''
710  T = 2
720  C$ = INKEY$(0)
730  IF C$ ='' '' GOTO 750
740  GOTO 720
750  LET S = ADVAL(1)/65520*800
760  PRINT TAB(0,30)''
                             ''
770  MOVE T,S
780  T = T + 1
790  IF POINT(T,S) = 1 THEN GOTO 820
800  DRAW T − 2,S
810  GOTO 780
820  MOVE T,S
830  DRAW T,0
840  PRINT TAB(0,0);''mmol. dm − 3''
850  PRINT TAB(0,L);T/100
860  L = L + 1
870  GOTO 690
880  END
```

A description of the program is presented below.

Line 40–140
Setting up routine.
Instructions for the experimenter are printed on the screen telling them to adjust the gain control on the signal conditioning unit. Line 80 takes in signals from the equipment and divides the number by 65520 then multiplies by 800 to bring it within the range of the Y co-ordinate of the screen graphics. This value of the signal is printed on the screen (line 90) and the operator is asked to adjust the gain control until the maximum (full scale deflection) output from the spectrophotometer is registered as 800 on the screen. The 800 corresponds to the Y co-ordinate value of an absorbance of 1.0 which is the maximum absorbance value on the graph drawn later in this program. Lines 100 and 110 slow down the sampling rate enough to see the numbers on the screen clearly. Lines 120 and 130 allow the experimenter to proceed by pressing the space bar once adjustment is complete.

Lines 150–460
Graph construction.
This has been fully explained in Chapter 5 on data presentation.

Lines 470–670
Plotting the graph of absorbance vs concentration of standard.
The first section (lines 470–600) consists of a series of prompts to tell the

experimenter when to type in the concentration of the standard, when to place the standard in the spectrophotometer and when to press the space bar to take a reading of absorbance.

Lines 610–670 is the plotting routine, an explanation of which can be found in Chapter 5.

Lines 680–880

Finding the concentration(s) of the unknown(s).

Following the prompt in Line 690, the computer accepts a signal from the equipment of the absorbance of the unknown. Then using the POINT statement in line 790 each pixel is examined in a horizontal line from that absorbance value on the axis until a lit pixel is recognized i.e. when the line joining the plotted points is reached. Line 850 then mathematically manipulates that value of T to calculate the glucose concentration. A line is drawn following the POINT statement (line 790) and a further line is drawn vertically downwards to the X axis from the position of intersection on the curve. The line has to be drawn behind the POINT statement (T − 2 in line 800) since the DRAW statement will light the pixels. The principle of this method is described in Chapter 6.

Appendix 1

Direct Input to the BBC Microcomputer's A–D Converter
The following items are required.

A 15 way 'D' plug (R.S. Components Ltd. Cat. No. 466–185)
A 15 way 'D' plug hood (R.S. Components Ltd. Cat. No. 469–572)
A 220 ohm resistor (R.S. Components Ltd. Cat. No. 132–337)
A BZX85 series 4.7 volt Zener diode (R.S. Components Ltd. Cat. No. 283–003)
Two insulated wires (preferably one red and one black)
A plug or plugs to connect to the output from the equipment you are interfacing to.

(i) Shorten the leads on the zener diode and resistor to approximately 8 mm long on either side and strip the ends of the two insulated wires.

(ii) Solder the two wires, the resistor and zener diode onto the rear of the 15 way 'D' plug as shown in Fig. A.1 below.

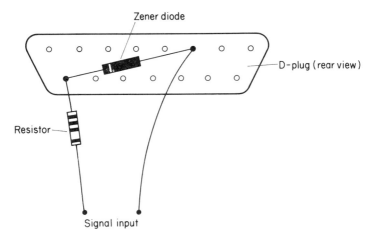

Fig. A.1 Connections to the 'analogue-in' port D-plug showing the protective zener diode and resistor.

(iii) Solder the plug(s) to connect to the appropriate piece of equipment onto the other end of the red and black wires.

(iv) Fit the 15 way 'D' plug hood to the D plug.

If the equipment is then connected to the 'analogue in' socket on the rear of the microcomputer using the lead just produced, the input can be tested using the following program.

```
10  LET B = ADVAL(1)
20  PRINT B
30  LET C = INKEY(100)
40  GOTO 10
```

To stop the program running press the ESCAPE key.

The zener diode and resistor have been inserted to protect the microcomputer from an excessive voltage and may be omitted; however this omission may risk damaging the computer if care is not taken with the voltage fed into the A–D converter. Many pieces of equipment will give voltage surges on switching ranges and when switched on or off.

Appendix 2

The Use of Osbyte Calls to Operate the User Port

In chapter 7 it was suggested that it is possible to control the User port by reading from or writing to memory locations 65122 and 65120 using the ? (query) instruction. However this is only possible on a standard 32K model B microcomputer, as on models with a second processor the organisation of the computer's memory is altered. Thus although in an uprated BBC microcomputer it may still be possible to read from and write to memory locations 65122 and 65120 they do not represent the User port. Consequently if this different type of BBC microcomputer is to be used, either now or in the future, it will be necessary to rewrite programs involving User port control.

To overcome this and other related problems, Acorn, the manufacturers of the BBC microcomputer, have a range of 'OSBYTE' calls built into the machine. These OSBYTE calls are principally designed for use in machine code programs and allow the programmer to perform a series of different functions, with the same OSBYTE call performing the same function whichever type of BBC microcomputer is being used.

Some of these OSBYTE calls can also be used with the *FX instructions. This is possible when programming the User port for output control, but not possible for input to the computer via the User port. To program the User port for output the *FX151 instruction followed by two numbers is used. The first of these two numbers may be either the number 98 or 96 which relate to writing to the data direction register (the equivalent of ?65122) or the input/output register (the equivalent of ?65120) respectively. The second number following the *FX151 instruction is the number which is to be entered into the data direction register or input/output register.

Thus to write to the data direction register:

        10 *FX151,98,255

is the equivalent of ?65122 = 255, and

        10 *FX151,98,0

is the equivalent of ?65122 = 0.

For the input/output register:

        20 *FX151,96,255

is the equivalent of ?65120 = 255, and

    30 *FX151,96,0

is the equivalent of ?65120 = 0

Thus the program on page 63 for switching on a relay connected to a relay connected to a relay line 0 for 5 seconds could be rewritten as follows:

```
1000 *FX151,98,255
1010 *FX151,96,1
1020 LET B = TIME
1030 REPEAT UNTIL TIME = B + 500
1040 *FX151,96,0
```

For input from the User port it is necessary to use the following piece of program:

```
10 *FX151,98,0
20 A% = 150
30 X% = 96
40 R = USR(65524)
50 PRINT ~ R
60 GOTO 20
```

In this program line 10 sets the data direction register for input and lines 20–40 are the equivalent of LET R = ?65120. If the input to the User port is changed whilst this program is running the number printed on the VDU screen will change (further details on the use of the USR instruction can be found in the BBC User Guide.) Unfortunately the number printed by this program is rather complex and difficult to decipher. Possibly the easiest way to determine the value obtained from the User port is to ensure that the first time the User port is read the input is at a set level and the value read therefrom is used in subsequent comparisons. The program below illustrates such a determination:

```
10 CLS
20 *FX151,98,0
30 PRINT;"ENSURE THAT ALL DATA LINES ARE SET TO 0
   VOLTS"
40 LET Z = INKEY(10000)
50 A% = 150
60 X% = 96
70 Q = USR(65524)
```

```
 80  CLS
 90  PRINT;"NOW APPLY YOUR SIGNAL TO THE DATA LINES"
100  LET Z = INKEY(10000)
110  A% = 150
120  X% = 96
130  R = USR(65524)
140  PRINT (R − Q)/65536
150  LET P = TIME
160  REPEAT UNTIL TIME = P + 50
170  GOTO 110
```

Appendix 3

Computer Programs which Illustrate the Storage of Numbers in a Binary Form in one (Eight Bit) Byte

In Chapter 3 the storage of numbers in one (eight bit) byte is described. This is a quite difficult concept to comprehend, and unfortunately a full and complete understanding is necessary before it is possible to maximize usage of certain areas of programming and interfacing (for example output control). To help in understanding data storage in a byte the following two programs have been devised.

The first program demonstrates how a number (0–255) entered into a byte is stored in terms of a bit pattern.

```
 10  MODE7
 20  PRINT TAB(5,3);"ENTER YOUR VALUE"
 30  PRINT TAB(22,3);"      "
 40  INPUT TAB(22,3) A
 50  PRINT TAB(34,8);"      "
 60  PRINT TAB(32,8);" = ";A
 70  LET B = 256
 80  FOR N = 7 TO 0 STEP − 1
 90  PRINT TAB(N*4,5);"BIT"
100  PRINT TAB((N*4 + 1),6);N
110  PRINT TAB((N*4 + 1),8);(A MOD B) DIV (B/2)
120  LET B = B/2
130  NEXT N
140  GOTO 30
```

In the second program it is possible to enter either a '0' or '1' into each bit in the byte to determine the number that corresponds to a certain bit pattern.

```
 10  MODE7
 20  D = 0
 30  PRINT TAB(1,3);"ENTER THE VALUE IN BIT"
 40  FOR N = 0 TO 7
 50  PRINT TAB(N*4,5);"BIT"
 60  PRINT TAB(N*4 + 1,6);N
 70  NEXT N
 80  FOR N = 0 TO 7
```

```
 90  PRINT TAB(24,3);N
100  INPUT TAB(26,3);A
110  PRINT TAB(26,3);''      ''
120  IF A > 1 THEN GOTO 90
130  PRINT TAB(N*4 + 1,8);A
140  IF A = 1 THEN LET D = D + 2^N
150  NEXT N
160  PRINT TAB(32,8);'' = '';D
170  PRINT TAB(1,3);''                                    ''
180  VDU 23;8202;0;0;0;
190  PRINT TAB(0,15);''PRESS THE SPACE BAR TO START
     AGAIN''
200  LET C$ = INKEY$(20000)
210  IF C$ = ''   '' THEN GOTO 10
220  GOTO 190
```

Appendix 4

The Use of Hexadecimal Notation

In the majority of situations it is possible to use conventional (decimal) numbers when programming. However there are situations in which it is important to have an understanding of hexadecimal notation. These situations are:

(i) When consulting publications which present their numbers in hexadecimal notation (for example: the BBC User Guide describes the addressing of memory locations in hexadecimal).

(ii) When saving to or loading from disc using the * SAVE and * LOAD commands. (See Chapter 4)

Hexadecimal notation uses the base 16 unlike decimal (base 10) and binary (base 2). This obviously needs extra digits for numbers from 10 to 15. The letters A to F are used for this as follows:

Decimal	Hexadecimal
1	1
2	2
3	3
4	4
5	5
6	6
7	7
8	8
9	9
10	A
11	B
12	C
13	D
14	E
15	F
16	10

It is not necessary to describe in detail how to convert decimal numbers into their hexadecimal counterpart since the computer can do this for you.

The BASIC language only understands a number to be in hexadecimal if & (called ampersand) appears before it, thus:

&2000 is considered as a hex number.

To convert this to decimal simply type into the computer:

PRINT &2000

which will return with the decimal equivalent 8192.

Decimal numbers can be converted to hexadecimal by entering:

PRINT ~ 8192

which will return with 2000.

When addressing memory using the indirection operators, memory locations may be addressed using either decimal or hexadecimal notation but if hexadecimal is used the number must be prefixed with &.

Appendix 5

A List of Equipment and Software Suppliers

The list below indicates a range of suppliers who sell equipment and software which is suitable for use in computer interfacing. This list is not intended to be comprehensive and it may be possible to obtain more suitable and/or cheaper equipment from other suppliers.

ACK Data,
21, Salcombe Drive,
Redhill,
Nottingham,
NG5 8JF,
U.K.

Business & Industrial Centre,
Lancashire Polytechnic,
Preston,
PR1 2TQ,
U.K.

Data Harvest,
28, Lake Street,
Leighton Buzzard,
Bedfordshire,
LU7 8RX,
U.K.

Griffin and George,
Head Office,
Ealing Road,
Alperton,
Wembley,
Middlesex,
HA0 1HJ,
U.K.

Palmer Bioscience,
Harbour Estate,
Sheerness,
Kent,
ME12 1RZ,
U.K.

Phillip Harris Ltd.,
Lynn Lane,
Shenstone,
Staffs.,
WS14 0EE,
U.K.

R.S. Components Ltd.,
P.O. Box 99,
Corby,
Northants,
NN17 9RS,
U.K.

Unilab Ltd.,
Clarendon Road,
Blackburn,
BB1 9TA,
U.K.

Index